PHYSICS RESEARCH AND TECHNOLOGY

A CLOSER LOOK AT BIOMECHANICS

PHYSICS RESEARCH AND TECHNOLOGY

Additional books and e-books in this series can be found
on Nova's website under the Series tab.

PHYSICS RESEARCH AND TECHNOLOGY

A CLOSER LOOK AT BIOMECHANICS

DANIELA FURST
EDITOR

Copyright © 2019 by Nova Science Publishers, Inc.

All rights reserved. No part of this book may be reproduced, stored in a retrieval system or transmitted in any form or by any means: electronic, electrostatic, magnetic, tape, mechanical photocopying, recording or otherwise without the written permission of the Publisher.

We have partnered with Copyright Clearance Center to make it easy for you to obtain permissions to reuse content from this publication. Simply navigate to this publication's page on Nova's website and locate the "Get Permission" button below the title description. This button is linked directly to the title's permission page on copyright.com. Alternatively, you can visit copyright.com and search by title, ISBN, or ISSN.

For further questions about using the service on copyright.com, please contact:
Copyright Clearance Center
Phone: +1-(978) 750-8400　　　Fax: +1-(978) 750-4470　　　E-mail: info@copyright.com

NOTICE TO THE READER

The Publisher has taken reasonable care in the preparation of this book, but makes no expressed or implied warranty of any kind and assumes no responsibility for any errors or omissions. No liability is assumed for incidental or consequential damages in connection with or arising out of information contained in this book. The Publisher shall not be liable for any special, consequential, or exemplary damages resulting, in whole or in part, from the readers' use of, or reliance upon, this material. Any parts of this book based on government reports are so indicated and copyright is claimed for those parts to the extent applicable to compilations of such works.

Independent verification should be sought for any data, advice or recommendations contained in this book. In addition, no responsibility is assumed by the Publisher for any injury and/or damage to persons or property arising from any methods, products, instructions, ideas or otherwise contained in this publication.

This publication is designed to provide accurate and authoritative information with regard to the subject matter covered herein. It is sold with the clear understanding that the Publisher is not engaged in rendering legal or any other professional services. If legal or any other expert assistance is required, the services of a competent person should be sought. FROM A DECLARATION OF PARTICIPANTS JOINTLY ADOPTED BY A COMMITTEE OF THE AMERICAN BAR ASSOCIATION AND A COMMITTEE OF PUBLISHERS.

Additional color graphics may be available in the e-book version of this book.

Library of Congress Cataloging-in-Publication Data

ISBN: 978-1-53615-866-3
Library of Congress Control Number:2019944891

Published by Nova Science Publishers, Inc. † New York

CONTENTS

Preface vii

Chapter 1 Biomechanical Testing of a Silicone Based
Elastomeric Augmentation Material in the Spine 1
*Werner Schmoelz, Javier M. Duart,
Luis Alvarez Galovich and Alexander Keiler*

Chapter 2 Biomechanical Behaviour of Traslational
Dynamic Cervical Plates.
Part 1: Finite Element Model 19
*Javier M. Duart, Werner Schmoelz,
Carlos Atienza, Ignacio Bermejo,
Darrel S. Brodke and Julio V. Duart*

Chapter 3 Biomechanical Behaviour of Traslational
Dynamic Cervical Plates. Part 2:
Biomechanical Essays 39
*Javier M. Duart, Werner Schmoelz,
Carlos Atienza, Ignacio Bermejo,
Darrel S. Brodke, Tobias Pitzen and
Julio V. Duart*

Chapter 4	A Biomechanical Study on Correlation between Laterality and Walking Asymmetry *Kadek Heri Sanjaya*	55
Chapter 5	Geometry and Inertia of the Human Body and Their Sport Applications *Wlodzimierz Stefan Erdmann*	99
Chapter 6	Clinical Application of Pedobarography *Arno Frigg*	133
Index		161
Related Nova Publications		167

Preface

The research presented in the opening chapter of A Closer Look at Biomechanics discusses the use of bone cements, and tests how a novel bone cement, medical grade two-component injectable polymer on silicone basis, can be used.

The second chapter demonstrates that the use of finite element modeling to simulate static and dynamic behavior in an anterior cervical plate design shows that load transmission is superior when the plate works dynamically.

The third chapter continues to examine the purpose of simulate static and dynamic behavior with the same anterior cervical plate design in two different clinical scenarios: in the immediate postoperative state and after simulated graft subsidence by means of biomechanical assays.

There are contradictory results from previous studies on the effects of laterality on walking, such as the existence of symmetry or asymmetry as well as the role of the dominant leg. Thus, the effects of laterality on walking asymmetry during walking on a treadmill is examined in this compilation.

The penultimate chapter discusses the localization of the body's center of mass and how that helps in the analyses of sport technique, while information on moment of inertia helps in explaining body angular movements.

The final chapter aims to show how the large number of pedobarographic parameters, which vary from 72 to 198 per foot, can be aggregated into a single indicative parameter: the Relative Midfoot Index. This indicates that clinicians do not have to analyze hundreds of pedobarographic parameters in order to reach a meaningful interpretation.

Chapter 1 - Augmentation with Polymethylmethacrylate (PMMA) cement is widely used for the treatment of vertebral fractures and to improve pedicle screw anchorage. Although PMMA has been used for many years as an augmentation-material in surgery, the application is still not without any negative side effects. For example, subsequent fractures after vertebroplasty are a common complication, since injection of a rigid material such as PMMA into the soft surrounding of an osteoporotic trabecular bone can lead to high interface stress, which causes interface failure at the cement-trabecular bone junction. A self-curing silicone based elastomeric material with mechanical properties closer to the trabecular bone structure might be an alternative material to PMMA. The present chapter gives an overview of in vitro experiments investigating the biomechanical properties of an elastomeric silicone material for the application in vertebroplasty procedures and for the Augmentation of pedicle screws. For vertebroplasty procedures two different injection volumes and materials were applied and the stiffness of the augmented vertebrae was compared. For the pedicle screw augmentation, the silicone based elastomer was applied with two different augmentation techniques and the screw anchorage was compared to the anchorage of pedicle screws conventionally in situ augmented with PMMA. Vertebroplasty procedures showed a reduced stiffness for the vertebrae treated with the silicone based elastomer compared to vertebrae augmented with PMMA. For pedicle screw augmentation, the silicone based elastomer showed a comparable or even superior number of load cycles until loosening compared to pedicle screws augmented in situ with PMMA, depending on the applied augmentation technique.

Chapter 2 - Fusion following anterior cervical decompressive procedures (such as discectomy and corpectomy) frequently involves bone and osteosynthesis with a plate in the case of corpectomies, where either a structural graft (from iliac crest or fibula) or a containing bone cylinder mesh are secured with an anterior plate to avoid extrusion and increase fusion chance. Plate design has evolved from unconstrained static to dynamic or semiconstrained plates, which in turn can be either rotational or traslational, with uni or bidirectional dynamicity. Apart from the clinical studies in the literature supporting dynamic plates, some biomechanical studies also favour them in comparison with static plates regarding load transmission which is thought to foster graft integration; nevertheless, hardly any study addresses the same plate design working under static and dynamic configuration, in order to discard material or geometrical properties which could explain part of the superior results both in vivo and in vitro scenarios. This study was arranged, to assess the biomechanical behaviour with finite element models, which are computed -based structures combining morphological and functional characteristics, with the ability to simulate both in vivo and in vitro situations. The purpose of this research work was to simulate static and dynamic behavior with the same anterior cervical plate design and in two different clinical scenarios, both in the immediate postoperative and after simulated graft subsidence (four simulated clinical scenarios in all) by means of finite element modelling, mimicking in vivo situations. Results show that load transmission is superior when the plate works dynamically, particularly after shortening of the graft so dynamic plates confer biomechanical advantages by improving transfer load and adapting to graft shortening. Chapter 3 - Anterior cervical decompressive procedures (such as discectomy and corpectomy) are usually followed by reconstruction for fusion, frequently involving bone and osteosynthesis with a plate in the case of corpectomies, where either a structural graft (from iliac crest or fibula) or a containing bone cylinder mesh are secured with an anterior plate to avoid extrusion and increase fusion chance. Plate design has evolved from unconstrained static to dynamic or semiconstrained plates, which in turn can be either rotational or traslational, with uni- or bidirectional dynamicity. Apart from the clinical studies in the literature supporting dynamic plates, some biomechanical studies also favour them in comparison with static plates regarding load transmission which is thought to foster graft integration; nevertheless, hardly any study addresses the same plate design working under static and dynamic configuration, in order to discard material or geometrical properties which could explain part of the superior results both *in vivo* and *in vitro* scenarios; for this reason, this study was arranged, to assess the biomechanical behaviour with mechanical essays. The purpose of this research work was to simulate static and dynamic behavior with the same anterior cervical plate design and in two different clinical scenarios, both in the immediate postoperative and after simulated graft subsidence (four simulated clinical scenarios in all) by means of biomechanical essays. Results show that load transmission is superior when the plate works dynamically, particularly after shortening of the graft so dynamic plates confer biomechanical advantages by improving transfer load and adapting to graft shortening.

Fusion following anterior cervical decompressive procedures (such as discectomy and corpectomy) frequently involves bone and osteosynthesis with a plate in the case of corpectomies, where either a structural graft (from iliac crest or fibula) or a containing bone cylinder mesh are secured with an anterior plate to avoid extrusion and increase fusion chance. Plate design has evolved from unconstrained static to dynamic or semiconstrained plates, which in turn can be either rotational or traslational, with uni or bidirectional dynamicity. Apart from the clinical studies in the literature supporting dynamic plates, some biomechanical studies also favour them in comparison with static plates regarding load transmission which is thought to foster graft integration; nevertheless, hardly any study addresses the same plate design working under static and dynamic configuration, in order to discard material or geometrical properties which could explain part of the superior results both in vivo and in vitro scenarios. This study was arranged, to assess the biomechanical behaviour with mechanical essays. The purpose of this research work was to simulate static and dynamic behavior with the same anterior cervical plate design and in two different clinical scenarios, both in the immediate postoperative and after simulated graft subsidence (four simulated clinical scenarios in all) by means of biomechanical essays, mimicking in vivo situations. Results show that load transmission is superior when the plate works dynamically, particularly after shortening of the graft so dynamic plates confer biomechanical advantages by improving transfer load and adapting to graft shortening.

Chapter 4 - The development of laterality in humans has been associated with the evolution of bipedalism. The most observable laterality is the handedness, where around 85 to 90% of the population is right-handed. Another laterality features measured in this study is the footedness. The correlation between the handedness and footedness is unclear, especially among the left-handers. There are contradictory results from the previous studies on the effects of laterality on walking, such as the existence of symmetry or asymmetry as well as the role of the dominant leg. The effects of walking speed on walking symmetry are also not clearly understood. This article discusses the effects of laterality on walking asymmetry during walking on a treadmill. Participants of the walking experiment were seventeen healthy young adult males (11 right-handers and right-footers and 6 left-handers and mixed-footers, measured by Waterloo Handedness and Footedness Questionnaire). In the recorded anthropometry data, the right-handers showed acromion height discrepancy whereas left-handers showed the trochanteric height discrepancy. Both groups of participants showed biceps and lower thigh circumference discrepancy. Participants walked on a treadmill at 1.5, 3, and 4 km/h. Bilateral muscle activation was measured from the tibialis anterior, soleus, and lumbar erector spinae. Foot pressure sensors were attached bilaterally on five points of foot sole: big toe, first, third, and fifth metatarsals, and calcaneus. Cross-correlation function (CCF) analysis was employed to analyze any paired root mean square (RMS) of each bilateral muscle activation and foot pressure signals, which yielded CCF coefficient and time lag during one gait cycle to measure the symmetry. In general, the authors' experiment showed that the left-handers had a greater asymmetry represented by lower CCF coefficient and longer time lag. In a further stage, CCF coefficient and time lag correlation with anthropometric data discrepancy were analyzed. The lower thigh circumference discrepancy, which ironically more correlated with handedness than footedness, was found to have the most consistent effects on the asymmetrical characteristics during walking as its effects were observed at all velocities measured. The greater asymmetry in left-handers probably related to their disadvantages in population study such as shorter life expectancy, higher safety risks, as well as musculoskeletal and sensory disorders. However, to what degree the disadvantages are embodied in physiological processes remain a big question necessary to be investigated in future studies with more appropriate experimental methods.

Chapter 5 - Human biomechanics encompasses body build, forces acting on it, and the results of this action, such as strain, deformation, and movement. Human body morphology can be described from the points of view of body structure, biomaterials, body construction, geometry, and inertia. There are many approaches for obtaining human body geometry and inertia. Scientists have used cadavers, body models, and living people. For external and internal geometry data, the following methods are utilized: direct measurements of the human body (anthropometry), photogrammetry, computerized tomography, and nuclear magnetic resonance. Inertial data are obtained most often by a) densitometry of tissues, body parts, or the whole body; b) mechanical or electronic scales for body mass; c) mechanical lever, photogrammetry for localization of body center of mass; or d) quick release, turntable, calculation methods for moment of inertia. There are several applications within athletic activity for human body geometric and inertial data. Athletic talent identification must take into account the dimensions of the body and their possible development within the course of the athlete's adolescence and youth. Athletic equipment must be built according to body dimensions of the members of current society. Many athletic disciplines are divided onto categories based on body mass (e.g., weight lifting, fighting sports). Localization of the body's center of mass helps in analyses of sport technique, while information on moment of inertia helps in explaining body angular movements.

Chapter 6 - Pedobarographic is widely used in research, but its application in clinical practice is still at the beginning. There are several reasons for this: (1) The installation of the equipment is time-consuming, which is a strain on the busy clinical schedule. (2) The equipment is relatively expensive, thereby pushing up costs. (3) It is often difficult to draw meaningful clinical conclusions from the large amount of data that pedobarographic exams provide. (4) Laboratory measurements often provide only limited insight into load patterns that feet are exposed to in the daily life of a patient. These limitations notwithstanding, pedobarography is becoming an increasingly useful and important instrument in the clinical context. In this article the authors show how the large number of pedobarographic parameters, which varies from 72 to 198 per foot, can be aggregated into a single indicative parameter, the so-called Relative Midfoot Index (RMI). This enables that clinicians do not have to analyze hundreds of pedobarographic parameters and instead can reach a meaningful interpretation by focusing only on the RMI, possibly combined with a visual interpretation of force/pressure time graphs in which healthy subjects show a triphasic force-time curve while diseased subjects show a biphasic force-time curve with a flatter midfoot depression. The authors therefore recommend a standardization of the reporting of pedobarographic outputs in terms of the RMI as well as the Maximal Force (as representation of load) and the contact times (as representation of rollover) in only three areas (hind-, mid- and forefoot). The article goes on to present clinical results obtained through the use of pedobarophy. (1) The authors show that contralateral feet cannot be used as a control group for a study of diseased feet because the gait of contralateral feet is significantly different from the gait of healthy feet. (2) Total ankle replacement shows no advantages over ankle arthrodesis when measured in shoes. This is clinically important because the authors almost always wear shoes, which makes barefoot measurements largely irrelevant. (3) Pedobarographic methods show that hindfoot alignment and the translation of radiological alignment into pedobarographic loading is crucial. This has the important clinical consequence that ankle arthrodesis have to be positioned in an angle of about 10° of valgus and that the standard *in situ* fusions of ankle osteoarthritis, which is typically in varus, yields inferior outcomes. Finally, the authors describe the new "third generation" pedobarographic systems and their potential in future research. First generation systems were pressure mats, which are nowadays available in commercial sports and shoe stores. Second generation systems were insole-sensors that measured foot-shoe interface pressures attached to cables in the laboratory. The new third generation systems are insoles with sensors and integrated or external data storage as well as power supply, making it possible to measure the foot-shoe interface over extended periods of time (days and weeks). Data are transmitted via Bluetooth, which makes cables unnecessary.

In: A Closer Look at Biomechanics ISBN: 978-1-53615-866-3
Editor: Daniela Furst © 2019 Nova Science Publishers, Inc.

Chapter 1

BIOMECHANICAL TESTING OF A SILICONE BASED ELASTOMERIC AUGMENTATION MATERIAL IN THE SPINE

Werner Schmoelz[1],*, PhD, Javier M. Duart[2], MD,
Luis Alvarez Galovich[3], MD
and Alexander Keiler[1], MD, PhD

[1]Departament of Trauma Surgery, Medical University of Innsbruck,
Innsbruck, Austria
[2]Spine and Neurosurgical Services, General Hospital, Valencia, Spain,
Departament de Cirurgia, Universitat Autónoma de Barcelona (UAB),
Barcelona, Spain
[3]Spine Service, Fundacion Jimenez Diaz University Hospital,
Madrid, Spain

ABSTRACT

Augmentation with Polymethylmethacrylat (PMMA) cement is widely used for the treatment of vertebral fractures and to improve

---

* Corresponding Author's E-mail: werner.schmoelz@i-med.ac.at.

pedicle screw anchorage. Although PMMA has been used for many years as an augmentation-material in surgery, the application is still not without any negative side effects. For example, subsequent fractures after vertebroplasty are a common complication, since injection of a rigid material such as PMMA into the soft surrounding of an osteoporotic trabecular bone can lead to high interface stress, which causes interface failure at the cement-trabecular bone junction.

A self-curing silicone based elastomeric material with mechanical properties closer to the trabecular bone structure might be an alternative material to PMMA. The present chapter gives an overview of in vitro experiments investigating the biomechanical properties of an elastomeric silicone material for the application in vertebroplasty procedures and for the Augmentation of pedicle screws.

For vertebroplasty procedures two different injection volumes and materials were applied and the stiffness of the augmented vertebrae was compared. For the pedicle screw augmentation, the silicone based elastomer was applied with two different augmentation techniques and the screw anchorage was compared to the anchorage of pedicle screws conventionally in situ augmented with PMMA.

Vertebroplasty procedures showed a reduced stiffness for the vertebrae treated with the silicone based elastomer compared to vertebrae augmented with PMMA. For pedicle screw augmentation, the silicone based elastomer showed a comparable or even superior number of load cycles until loosening compared to pedicle screws augmented in situ with PMMA, depending on the applied augmentation technique.

Keywords: elastoplasty, vertebroplasty, pedicle screw anchorage, augmentation, PMMA, VK100, silicone based elastomer

INTRODUCTION

In geriatric patients with reduced bone quality fracture, stabilization and reduction can be challenging to the surgeon. Therefore, bone cements are used to enhance the anchorage of pedicle screws and prevent loosening of the screws after repetitive loading or to stabilize a fracture with the vertebroplasty or kyphoplasty technique. Various types of bone cements with varying advantages and disadvantages are available to the surgeon in the operating room. They can be divided into two main groups: ceramics and polymers. All calcium based bone cements can be allocated to the ceramic group, whereas Polymethylmethacrylate (PMMA) and silicone based elastomers are both polymers. Calcium based bone cements are usually resorbable, while polymer based bone cements are typically non-resorbable. In Europe, PMMA bone cements are widely used for fracture stabilization and implant augmentation in the spine. Despite its use for many years, negative side effects and drawbacks are reported in the literature for the application of PMMA cement in the spine. Raised concerns in the literature range from thermal necrosis due to its exothermic curing behaviour [1-3], limited time frame for processing [4], lack of osteoconductivity [5] to toxic properties of the evaporated monomer during curing [6-9].

Furthermore, the mechanical properties of the trabecular bone of geriatric patients differs substantially from the characteristics/features of the injected PMMA bone cement, which is much stiffer and rigid than the trabecular bone structure. These stiffness differences cause high interface stresses and bear the risk of interface failure at the cement-PMMA boundary [10]. Additionally, several studies report an increase in vertebral stiffness after PMMA injection, which may cause alterations in the kinetic chain and load transfer of the spinal column resulting in an increased fracture risk and an increased incidence of adjacent fractures [10-15].

Due to these shortcomings of available bone cements, research has focused on the development of novel bone cements. One possible approach might be a medical grade two-component injectable polymer on silicone basis. Silicones have expanded in the use of the medical field since the 1960s. They have non-exothermic curing properties and are intended to be osteoconductive and non-hazardous to the surrounding tissue [16, 17]. The bulk elastic modulus of the silicone based bone cement was engineered to be lower than PMMA and closer to the bulk modulus of the trabecular bone.

The present chapter summarizes experimental studies on functional in vitro testing of a novel silicone based bone cement. The mechanical effects of the silicone based elastomer for its use in vertebroplasty procedures as well as for the Augmentation of pedicle screws are studied and compared to PMMA based bone cements.

METHODS

In the subsequent sections, laboratory investigations of functional in vitro testing with an elastomeric silicone based bone cement (VK100, BONWRx, Lansing, MI, USA) are summarized. The summary comprises of previously published studies on the application of VK100 for vertebroplasty procedures [18] as well as for the Augmentation of pedicle screws to improve pedicle screw anchorage [16], supplemented by previously unpublished data.

Vertebroplasty Application

The in vitro study by Schulte et al. investigated the stiffness of fractured vertebrae augmented with PMMA or VK100 for two different injection volumes [18]. Vertebral compression fractures were created in 40 isolated thoracolumbar vertebrae in a servohydraulic material testing machine by compressing them in a specifically designed jig to 30% of the anterior vertebral body height (Figure 1). Prior to fracturing of the vertebrae, the volume of the vertebral bodies was measured and the bone mineral density (BMD) was determined using a qCT scan. According to their age, BMD and vertebral level, vertebrae were assigned to one of the four test groups. Using the vertebroplasty technique for both Augmentation materials (PMMA and VK100), two different volumes of the bone cement were injected (16 and 35% of intact vertebral body volume). With two materials and two filling grades, four groups were available for comparison (16PMMA, 35PMMA, 16VK100 and 35VK100).

After Augmentation, the same test setup for fracture creation was used for cyclic loading of the treated vertebrae (Figure 1). The cyclic loading protocol was comprised of 5000 load cycles (0.5Hz) ranging from 20 to 65% of the maximum load recorded during fracture creation. Among other parameters, the stiffness for the applied load range during cyclic loading was evaluated and normalized to the stiffness of the corresponding vertebra measured during fracture creation [18].

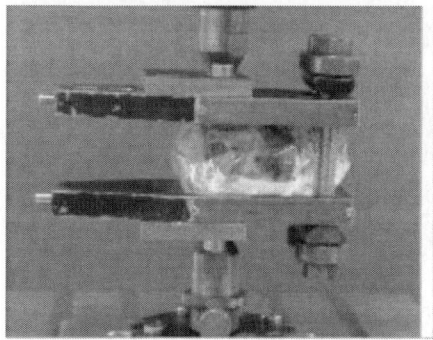 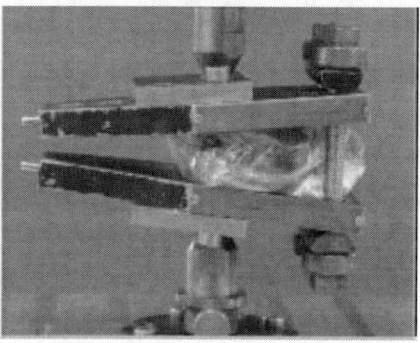

Figure 1. Setup for fracture creation and cyclic loading in a material testing machine with standardized load application in the ventral third of the vertebral both; via a ball and socket joint.

Augmentation of Pedicle Screws

Pedicle screws can be augmented with bone cement in different techniques [19, 20], which in general can be summarized in two different methods of cement application. Either "in situ" after placement of the pedicle screw using cannulated and fenestrated pedicle screws, or by inserting the pedicle screw in a pre-prepared and cement filled screw trajectory or balloon created cavity. A published study by Schmoelz et al. compared pedicle screw anchorage of in situ PMMA augmented screws to screws inserted in a balloon cavity filled with an elastomer based silicone (VK100) [16]. In a follow up study, the same application technique was used for both materials, and the anchorage of pedicle screws after in situ Augmentation with either PMMA or VK100 through cannulated and fenestrated screws was compared.

For both studies on pedicle screw anchorage, the same test setup and loading protocol was used. In the test setup, each pedicle screw was cyclically loaded in a material testing machine. Vertebrae were mounted on an x-y plane bearing table and axial loading in craniocaudal direction and was applied to the pedicle screws via a straight rod connected to a rotational axis with an offset lever arm of 15mm. By this setup, the pivoting point of the pedicle screw during cyclic loading was shifted to the middle of the vertebral pedicle. The applied cyclic loading protocol ranged initially from 50N in tension to 50N in compression (speed 5mm/sec) with an increase in compressive load of 5N every 100 load cycles for a total of 11000 cycles, which corresponds to 600N of compressive loading at the last cycle [16, 21-23]. Relative motion of the pedicle screw within the vertebral body was measured using a 3-D motion analysis system mounted to the pedicle screw head and to the base plate of the x-y table (Figure 2).

This setup allowed a pairwise left/right comparison of two Augmentation techniques or materials in the pedicles of one vertebra with comparable bone mineral density (BMD) and morphology for both screws. Post-testing data of the relative motion of each pedicle screw within the vertebral body was analysed to determine the number of load cycles and corresponding load magnitude until the loosening of each pedicle screw [16, 21-23].

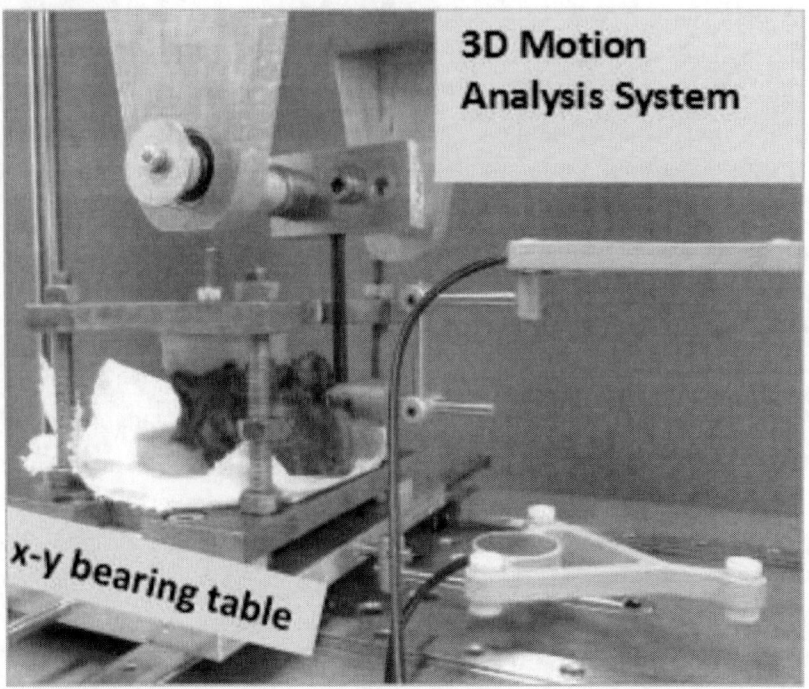

Figure 2. Test setup for cyclic loading of pedicle screws in craniocaudal direction with a 3D motion analysis system attached to the screw head and to the x-v bearing table.

In the initial study, ten lumbar vertebrae were used to compare the anchorage of in situ with PMMA augmented pedicle screws to the anchorage of pedicle screws inserted in VK100 filled voids created by the kyphoplasty technique [16]. Specimens had a mean age of 77.7 ±8.7 years and a mean BMD of 92.1 ±33.6 mg/cm^3. All right pedicles were instrumented with cannulated and fenestrated pedicle screws and in situ augmented with 2ml of PMMA cement, while left pedicle screws were inserted in a balloon created cavity filled with 3ml of VK100 [16].

In a follow up study, eight lumbar vertebrae were used to compare the anchorage of pedicle screws augmented in situ with PMMA to pedicle screws augmented in situ with VK100. Specimens had a mean age of 82.5 ±9 years and a mean BMD of 70.2 ±36.6 mg/cm^3. All right pedicles were instrumented with cannulated and fenestrated pedicle screws and augmented with 3 ml of VK100. All left pedicles were instrumented with cannulated and fenestrated pedicle screws and augmented with 2 ml of PMMA.

RESULTS

Vertebroplasty Application

The mean overall fracture loads of all groups was 3331N (SD 876N), ranging from 3009N to 3667N for the four groups. Wedge type fractures were created in all vertebrae [18] and the fracture loads between the groups were not significantly different (p=0.67).

The normalized stiffness during cyclic loading increased in all four groups and reached 361% and 304% in the 35% and 16% PMMA augmented groups, and 243% and 222% for the 35% and 16% VK100 augmented groups at the end of the cyclic loading [18]. Pooling the data for the material and the filling grade showed a significantly lower stiffness of the vertebrae augmented with VK100 compared to the augmentation with PMMA, independent of the filling grade (Figure 3).

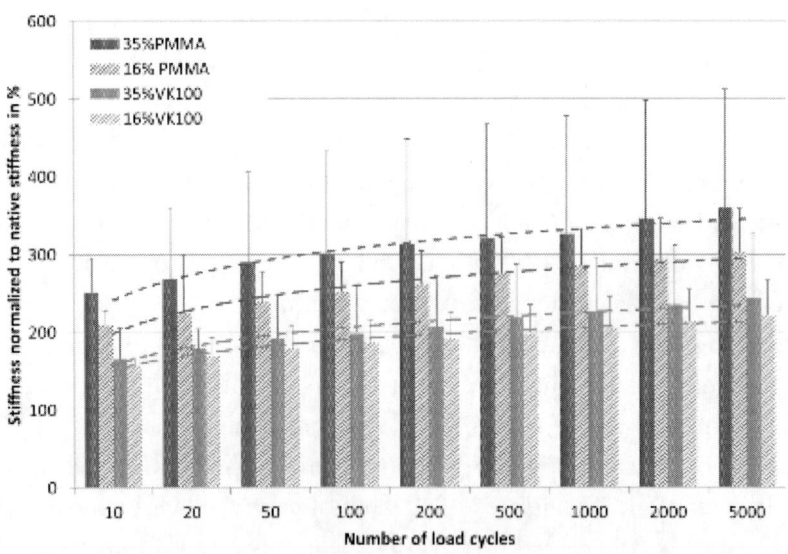

Figure 3. Normalized mean cyclic stiffness of the four test groups in the course of cyclic loading with logarithmic trend lines for the four test groups.

Augmentation of Pedicle Screws

In the initial study comparing in situ augmentation with PMMA to balloon cavities filled VK100 Augmentation, nine vertebrae with eighteen pedicle screws were available for data analysis [16]. In seven of the nine vertebrae, the balloon cavity VK100 augmented screws showed a higher number of load cycles and a higher load magnitude than the screws augmented in situ with PMMA (Figure 4).

The mean number of load cycles until loosening was detected to be summed up to 7401 ±1645 for in situ PMMA augmented screws and to 9824 ±1982 for the VK100 balloon cavity augmented screws (p=0.012). These numbers of load cycles corresponded to a mean load magnitude of 420 ±82N for the in situ PMMA augmented screws and to 542 ±99N for the balloon cavity VK100 augmented screws [16].

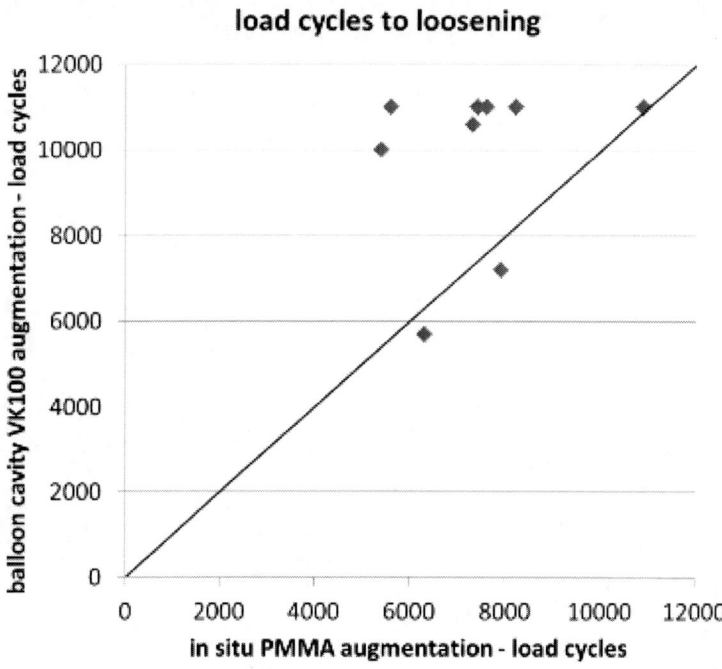

Figure 4. Number of load cycles until loosening of in situ PMMA augmented pedicle screws and of pedicle screws inserted in a balloon cavity filled with VK100. One dot represents one vertebra with one screw of each augmentation technique.

In the follow up study comparing the two materials (PMMA and VK100) for in situ screw Augmentation, all eight vertebrae and sixteen pedicle screws were available for data analysis. In four of the eight vertebrae, VK100 augmented pedicle screws reached a higher number of load cycles and a higher load magnitude, while in four vertebrae PMMA augmented screws reached a higher number of load cycles until loosening (Figure 5). The mean number of load cycles until loosening was detected to be summed up to 6084 ±1859 for in situ PMMA augmented screws and to 5321 ±2163 for the in situ VK100 augmented screws (p=0.53). This number of load cycles until loosening of the screws corresponded to a mean load magnitude of 351 ±92 N for the in situ PMMA augmented screws and to 313±109 N for the in situ VK100 augmented screws (p=0.53).

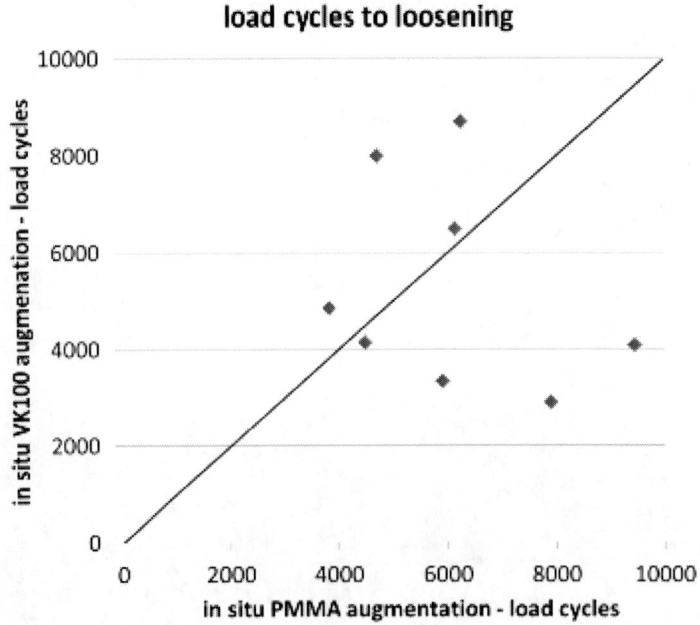

Figure 5. Number of load cycles until loosening of in situ PMMA and VK100 augmented pedicle screws. One dot repr esents one vertebra with one screw of each augmentation technique.

DISCUSSION

The conducted biomechanical studies revealed that the tested silicone based elastomeric material may be a viable option in the treatment of osteoporotic fractures and augmentation of pedicle screws [16, 18]. Biomechanically, the stiffness of the elastomeric material was designed to resemble the compound stiffness of trabecular bone and not the stiffness of a single trabecula. Therefore, the greatest benefit of the elastomeric material might be gained by applying the material in bulk. This can be achieved either with the kyphoplasty technique for the treatment of vertebral fractures or by creating a balloon cavity and filling it with the silicone elastomer prior to insertion of a pedicle screw [16, 18].

Vertebroplasty Application

After fracturing and stabilizing the vertebrae in the vertebroplasty technique, the biomechanical experiments showed for both augmentation materials a significantly higher stiffness than the corresponding native vertebrae. In comparison of the two materials, the stiffness of VK100 treated vertebrae was significantly lower than the vertebrae treated with PMMA. As expected, injecting more material resulted in a higher stiffness for both materials. While enlarging the augmentation volume from 16% to 35% had only a small stiffness increasing effect for VK100, it had a substantial effect on the vertebral stiffness for PMMA.

In a clinical study on predictive factors for adjacent vertebral fractures after vertebroplasty, Ahn et al. described the increase in stiffness of a PMMA treated vertebra as being a predictive factor for subsequent fractures of adjacent vertebrae [24]. Several in vitro and in silico studies also reported an increased stiffness after PMMA augmentation as a risk factor for subsequent adjacent fractures [10-15]. Therefore, research efforts focused on the development of new and modified augmentation materials with mechanical properties closer to the characteristics of trabecular bone structure. Among other inventions, this encompassed the development of low modulus PMMA cement, modified calcium phosphate-based cements and polymers. Biomechanical experiments of alternative materials with a lower modulus showed a reduction in fracture risk of adjacent vertebral compared to conventional PMMA [13, 25].

Augmentation of Pedicle Screws

Previous biomechanical studies on pedicle screw anchorage consistently reported a significant improvement in pedicel screw anchorage for varying augmentation techniques and materials compared to standard non-augmented pedicle screws [19, 20, 26, 27]. Therefore, the present biomechanical experiments focused on comparisons of different augmentation materials and techniques. Augmenting pedicle screws with the in situ technique using VK100 was feasible and pedicle screw anchorage was comparable to in situ PMMA augmentation, while pedicle screws inserted in the balloon created cavities filled with VK100. This significantly increased pedicle screw anchorage compared to in situ PMMA augmented pedicle screws.

The present test setup with cyclic loading of pedicle screws in craniocaudal direction is commonly applied in the literature to investigate screw anchorage. It is described to replicate the in vivo loading of pedicle screws much better than non-physiological pullout tests of pedicle screws [21-23, 27-29]. A stepwise or constantly increasing load magnitude for a cyclic loading protocol is well suitable to investigate implant anchorage in varying bone quality and for the varying stiffness of the instrumentation [20, 27, 30, 31].

Weiser et al. investigated the influence of BMD on pedicle screw anchorage for conventional non-augmented and in situ PMMA augmented pedicle screws. They reported a significant correlation of BMD and failure loads for both conventional non-augmented and PMMA augmented pedicle screws. This correlation of BMD with failure loads could be a reason for the observed difference in failure loads of the in situ augmented PMMA in the two studies. With a higher mean BMD in the first study, the mean failure loads were also higher than for the follow up study, which had a lower BMD.

Putting the failure loads for pedicle screws of the presented results in context with in vivo acting loads measured in patients with an instrumented internal fixator showed that failure loads for both augmentation techniques and materials were above the load magnitudes reported in patients during everyday activities [32, 33].

Clinical Application for Vertebral Augmentation

After functional biomechanical testing of the novel silicone based elastomeric material in the laboratory, it was applied in a clinical case series for vertebral augmentation. Clinical safety and feasibility studies for the use of VK100 in kyphoplasty and vertebroplasty procedures were carried out [34-37]. These studies concluded that augmentation with the VK100 is a safe and promising procedure with improvements in VAS and ODI after treatment comparable to PMMA augmentation. Cement leakage outside the vertebral body occurred in 6 to 17% of patients in kyphoplasty procedures and in 13% to 31% of patients in vertebroplasty procedures. These leakage rates occurring for VK100 vertebral augmentation are in the order of the leakage rates reported for PMMA cement in vertebroplasty and kyphoplasty procedures in the literature [38, 39]. However, the handling of the silicone based elastomeric material is different to the handling of PMMA. Therefore, augmentation with VK100 requires training to avoid leakages as reported by Urlings et al. who presented a first case of series of 12 patients with high incidences of perivertebral leakage and pulmonary embolism [40].

In order to investigate if the incidence of adjacent vertebral fractures after treatment with a less stiff augmentation material can be reduced, Bornemann et al. compared the rate of adjacent fractures occurring in patients after kyphoplasty procedures with either VK100 or PMMA cement. With a follow up of 12 months in 30 patients, they reported a trend towards less adjacent fractures and concluded that the number of patients as well as the follow up period should be increased and additional parameters, such as bone mineral density and the amount of injected filling material, should also be considered in the evaluation of the results.

REFERENCES

[1] Belkoff, S. M. and Molloy, S. (2003). Temperature measurement during polymerization of polymethylmethacrylate cement used for vertebroplasty. *Spine,* 28:1555-9.

[2] Jefferiss, C. D., Lee, A. J. and R. S. Ling (1975). Thermal aspects of self-curing polymethylmethacrylate. *The Journal of bone and joint surgery, British volume*, 57: 511-8.

[3] Stańczyk, M. and van Rietbergen, B. (2004). Thermal analysis of bone cement polymerisation at the cement-bone interface. *Journal of biomechanics*, 37:1803-10.

[4] Breusch, S. J. and Kühn, K. D. (2003). Knochenzemente auf Basis von Polymethylmethacrylat [Bone cements based on polymethylmethacrylate]. *Der Orthopäde*, 32:41-50.

[5] Freeman, M. A., Bradley, G. W. and Revell, P. A. (1982). Observations upon the interface between bone and polymethylmethacrylate cement. *The Journal of bone and joint surgery, British volume*, 64:489-93.

[6] Ciapetti, G., Granchi, D., Cenni, E., Savarino, L., Cavedagna, D., and Pizzoferrato, A. (2000). Cytotoxic effect of bone cements in HL-60 cells: distinction between apoptosis and necrosis. *Journal of biomedical materials research*, 52:338-45.

[7] Granchi, D., Stea, S., Ciapetti, G., Savarino, L., Cavedagna, D., and Pizzoferrato, A. (1995). In vitro effects of bone cements on the cell cycle of osteoblast-like cells. *Biomaterials*, 16:1187-92.

[8] Kalteis, T., Lüring, C., Gugler, G., Zysk, S., Caro, W., Handel, M. and J. Grifka (2004). Akute Gewebetoxizität von PMMA-Knochenzementen [Acute tissue toxicity of PMMA bone cements]. *Zeitschrift für Orthopädie und ihre Grenzgebiete*, 142: 666-72.

[9] Leggat P. A., Smith D. R. and Kedjarune U. (2009). Surgical applications of methyl methacrylate: a review of toxicity. *Arch Environ Occup Health*, 64(3):207-12.

[10] Kinzl, M., Boger, A., Zysset, P. K., and Pahr, D. H. (2012). The mechanical behavior of PMMA/bone specimens extracted from augmented vertebrae: a numerical study of interface properties, PMMA shrinkage and trabecular bone damage. *J Biomech*, 45(8):1478-84.

[11] Baroud, G., Heini, P., Nemes, J., Bohner, M., Ferguson, S. and Steffen, T. (2003). Biomechanical explanation of adjacent fractures following vertebroplasty. *Radiology*, 229(2):606-7; author reply 607-8.

[12] Berlemann, U., Ferguson, S. J., Nolte, L. P. and Heini, P. F. (2002). Adjacent vertebral failure after vertebroplasty. A biomechanical investigation. *J Bone Joint Surg Br,* 84(5):748-52.

[13] Boger, A., Heini, P., Windolf, M. and E. Schneider (2007). Adjacent vertebral failure after vertebroplasty: a biomechanical study of low-modulus PMMA cement. *Eur Spine J,* 16(12):2118-25.

[14] Polikeit, A., Nolte, L. P. and Ferguson, S. J. (2003). The effect of cement augmentation on the load transfer in an osteoporotic functional spinal unit: finite-element analysis. *Spine,* 28(10):991-6.

[15] Wilcox, R. K. (2006). The biomechanical effect of vertebroplasty on the adjacent vertebral body: a finite element study. *Proc Inst Mech Eng H,* 220(4):565-72.

[16] Schmoelz, W., Keiler, A., Konschake, M., Lindtner, R. A. and Gasbarrini, A. (2018). Effect of pedicle screw augmentation with a self-curing elastomeric material under craniocaudal cyclic loading: A cadaveric biomechanical study. *J Orthop Surg Res,* 13(1):251.

[17] Song, W., Seta, J., Eichler, M. K., Arts, J. J., Boszczyk, B. M., Markel, D.C., Gasbarrini, A. and Ren, W. (2018). Comparison of in vitro biocompatibility of silicone and polymethyl methacrylate during the curing phase of polymerization. *J Biomed Mater Res B Appl Biomater,* 106(7):2693-2699.

[18] Schulte, T. L., Keiler, A., Riechelmann, F., Lange, T. and Schmoelz, W (2013). Biomechanical comparison of vertebral augmentation with silicone and PMMA cement and two filling grades. *Eur Spine J,* 22(12):2695-701.

[19] Becker, S., Chavanne, A., Spitaler, R., Kropik, K., Aigner, N., Ogon, M. and Redl, H. (2008). Assessment of different screw augmentation techniques and screw designs in osteoporotic spines. *European spine journal,* 17: 1462-9.

[20] Bostelmann R., Keiler A., Steiger H. J., Scholz A., Cornelius J. F., Schmoelz W. (2017). Effect of augmentation techniques on the failure of pedicle screws under craniocaudal cyclic loading. *Eur Spine J,* 26(1):181-188.

[21] Schmoelz, W., Heinrichs C. H., Schmidt S., Pinera A. R., Tome-Bermejo F., Duart J. M., Bauer M., Galovich L. A. (2017). Timing of PMMA cement application for pedicle screw augmentation affects screw anchorage. *Eur Spine J,* 26(11): 2883-2890.

[22] Lindtner, R. A., Schmid, R., Nydegger, T., Konschake, M. and Schmoelz, W. (2018). Pedicle screw anchorage of carbon fiber-reinforced PEEK screws under cyclic loading. *Eur Spine J.* 27(8): 1775-1784.

[23] Spicher, A., Lindtner, R. A., Zimmermann, S., Stofferin, H. and Schmoelz, W. (2019). Ultrasound melted polymer sleeve for improved primary pedicle screw anchorage: A novel augmentation technique. *Clin Biomech,* 63:16-20.

[24] Ahn, Y., Lee, J. H., Lee, H. Y., Lee, S. H. and Keem, S. H. (2008). Predictive factors for subsequent vertebral fracture after percutaneous vertebroplasty. *J Neurosurg Spine,* 9(2):129-36.

[25] Nouda, S., Tomita, S., Kin, S., Kawahara. A. and Kinoshita, M. (2009). Adjacent vertebral body fracture following vertebroplasty with polymethylmethacrylate or calcium phosphate cement: biomechanical evaluation of the cadaveric spine. *Spine,* 34(24): 2613-8.

[26] Renner, S. M., Lim, T. H., Kim, W. J., Katolik, An, H. S. and Andersson, G. B. J. (2004). Augmentation of pedicle screw fixation strength using an injectable calcium phosphate cement as a function of injection timing and method. *Spine,* 29:E212-6.

[27] Kueny, R. A., Kolb, J. P., Lehmann, W., Puschel, K., Morlock, M. M. and Huber, G. (2014). Influence of the screw augmentation technique and a diameter increase on pedicle screw fixation in the osteoporotic spine: pullout versus fatigue testing. *Eur Spine J,* 23(10):2196-202.

[28] Weiser, L., Huber, G., Sellenschloh, K., Viezens, L., Puschel, K., Morlock, M. M. and Lehmann, W. (2018). Time to augment?! Impact of cement augmentation on pedicle screw fixation strength depending on bone mineral density. *Eur Spine J,* 27(8): 1964-1971.

[29] Tan, J. S., Bailey, C. S., Dvorak, M. F., Fisher, C. G., Cripton, P. A. and Oxland, T. R. (2007). Cement augmentation of vertebral screws enhances the interface strength between interbody device and vertebral body. *Spine,* 2(3):334-41.
[30] Unger, S., Erhart, S., Kralinger, F., Blauth, M. and Schmoelz, W. (2012). The effect of in situ augmentation on implant anchorage in proximal humeral head fractures. *Injury,* 43(10):1759-63.
[31] Windolf, M., Maza, E. R., Gueorguiev, B., Braunstein, V. and Schwieger, K. (2010). Treatment of distal humeral fractures using conventional implants. Biomechanical evaluation of a new implant configuration. *BMC musculoskeletal disorders*, 11:172.
[32] Rohlmann, A., Bergmann, G. and Graichen, F. (1999). Loads on internal spinal fixators measured in different body positions. *Eur Spine J,* 8(5): 354-9.
[33] Rohlmann, A., Graichen, F. and Bergmann, G. (2000). Influence of load carrying on loads in internal spinal fixators. *Journal of biomechanics,* 33: 1099-104.
[34] Bornemann, R., Rommelspacher, Y., Jansen, T. R., Sander, K., Wirtz, D. C. and Pflugmacher, R. (2016). Elastoplasty: A Silicon Polymer as a New Filling Material for Kyphoplasty in Comparison to PMMA. *Pain Physician*, 19(6): E885-92.
[35] Gasbarrini, A., Ghermandi, R., Akman, Y. E., Girolami, M. and Boriani, S. (2017). Elastoplasty as a promising novel technique: Vertebral augmentation with an elastic silicone-based polymer. *Acta Orthop Traumatol Turc*, 51(3): 209-214.
[36] Mauri, G., Nicosia, L., Sconfienza, L. M., Varano, G. M., Vigna, P. D., Bonomo, G., Orsi, F. and Anselmetti, G. C. (2018). Safety and results of image-guided vertebroplasty with elastomeric polymer material (elastoplasty). *Eur Radiol Exp,* 2(1): 31.
[37] Telera, S., Pompili, A., Crispo, F., Giovannetti, M., Pace, A., Villani, V., Fabi, A., Sperduti, I. and L. Raus (2018). Kyphoplasty with purified silicone VK100 (Elastoplasty) to treat spinal lytic lesions in cancer patients: A retrospective evaluation of 41 cases. *Clin Neurol Neurosurg,* 171:184-189.

[38] Hulme, P. A., J. Krebs, S. J. Ferguson, and Berlemann, U. (2006). Vertebroplasty and kyphoplasty: a systematic review of 69 clinical studies. *Spine* 31(17):1983-2001.

[39] Wang, H., Sribastav, S.S., Ye, C. Yang, J. Wang, H. Liu, and Z. Zheng (2015). Comparison of Percutaneous Vertebroplasty and Balloon Kyphoplasty for the Treatment of Single Level Vertebral Compression Fractures: A Meta-analysis of the Literature. *Pain Physician,* 18(3):209-22.

[40] Urlings, T. A. and van der Linden, E. (2013). Elastoplasty: first experience in 12 patients. *Cardiovasc Intervent Radiol,* 36(2):479-83.

Chapter 2

BIOMECHANICAL BEHAVIOUR OFTRASLATIONAL DYNAMIC CERVICAL PLATES. PART 1: FINITE ELEMENT MODEL

Javier M. Duart[1],*, Werner Schmoelz[1], Carlos Atienza[3], Ignacio Bermejo[3], Darrel S. Brodke[4] and Julio V. Duart[5]

[1]Spine and Neurosurgical Services, General Hospital, Valencia, Spain; Departament de Cirurgia, Universitat Autónoma de Barcelona (UAB), Barcelona, Spain
[2]Dept. of Trauma Surgery, Medical University of Innsbruck, Innsbruck, Austria
[3]Institute of Biomechanics of Valencia, Valencia, Spain
[4]Orthopedic Department, University of Utah, Salt Lake City, Utah
[5]Orthopedic Department, Hospital de Estella, Estella-Lizarra, Spain

ABSTRACT

Fusion following anterior cervical decompressive procedures (such as discectomy and corpectomy) frequently involves bone and osteosynthesis with a plate in the case of corpectomies, where either a

---

* Corresponding Author's E-mail: jduart@uv.es.

structural graft (from iliac crest or fibula) or a containing bone cylinder mesh are secured with an anterior plate to avoid extrusion and increase fusion chance. Plate design has evolved from unconstrained static to dynamic or semiconstrained plates, which in turn can be either rotational or traslational, with uni or bidirectional dynamicity. Apart from the clinical studies in the literature supporting dynamic plates, some biomechanical studies also favour them in comparison with static plates regarding load transmission which is thought to foster graft integration; nevertheless, hardly any study addresses the same plate design working under static and dynamic configuration, in order to discard material or geometrical properties which could explain part of the superior results both in vivo and in vitro scenarios.

This study was arranged, to assess the biomechanical behaviour with finite element models, which are computed -based structures combining morphological and functional characteristics, with the ability to simulate both in vivo and in vitro situations. The purpose of this research work was to simulate static and dynamic behavior with the same anterior cervical plate design and in two different clinical scenarios, both in the immediate postoperative and after simulated graft subsidence (four simulated clinical scenarios in all) by means of finite element modelling, mimicking in vivo situations. Results show that load transmission is superior when the plate works dynamically, particularly after shortening of the graft, so dynamic plates confer biomechanical advantages by improving transfer load and adapting to graft shortening.

Keywords: biomechanics, corpectomy, dynamic plate

INTRODUCTION

Anterior cervical plates are used for stabilization of the cervical vertebrae after decompressive procedures to enhance fusion; those which allow for movement being transmitted to the bone graft to foster union are called dynamic plates, and seem to offer some biomechanical and clinical advantages over static ones. In turn, dynamic can be divided into rotational or traslational (uni- or bidirectional), depending on their mechanism of action; bidirectional dynamic cervical seem to be superior at least in theory to the other types of plates.

The aim of this work is to analyze the biomechanical behaviour of the dynamic plates opposite to static ones; to avoid possible differences from the design or the material of the plates, the same plate design working in dynamic and static modes will be used integrated in the corpectomy model, and both the immediate postoperative situation as well as after simulate shortening of the graft during its integration during the fusion period.

Finite element models are mathematical models drawn by computer to integrate physical properties of each one of the parts (including their morphology) into mechanical properties as a group; they have been widely used for implant evaluation.

1. METHODS

The research group which stands out in the field of finite element modelling in the cervical spine region is the Neurosurgical Department of the Medical College of Wisconsin [1-9]. This group developed geometric models using images from X-rays or CT scan using an algorithm of edge detection. The meshing of the developed geometric was done using hexahedric elements of 8 nodes, but for cortical bone and vertebral endplates, using plate elements of 4 nodes, and for the modelling of ligaments, in which cable elements were used.

1.1. Material Characteristics

Given that the assigned mechanic properties to each material of the finite element model (bone, ligaments, etc.) are going to be based in data from other studies, a bibliographic search was performed. For the properties of the cortical and trabecular bone, data was taken from the bibliography related to other finite element models of the cervical spine (Table 1). The mechanic and geometric (section area) properties of the different ligaments of the model were obtained from the study of Yoganandan [10, 11], in which the geometric values of the ligaments were obtained from cervical spines taken from 8 human cadavers, while the data regarding the mechanical properties of the ligaments were obtained from

25 human cadavers. For the mechanic characterization of the bone graft, another search was performed [12-17] and the assigned mechanical properties taken for the study (Table 2).

Among all the articles found during these searches, none of them included the values of Young modulus (E) and Poisson coefficient, and none included mechanical evaluation of bone grafts when performed.

Table 1. Mechanic properties, employed in the FEM

bone	E (N/mm2)	Poisson coefficient
Cancellous	100	0'2
Cortical	12000	0'3
Cancellous of the posterior arch	6000	0'3
ligaments	**Total area (mm2)**	**E1 (Mpa)**
Anterior longitudinal	12.1	28.2
Posterior longitudinal	14.7	23
Yellow	48.9	3.5
Interspinous	13.4	5
Supraspinous	20	

Table 2. Mechanical properties of the bone graft employed for the FEM

	E (Mpa)	Poisson coefficient	Density (g/cm3)
Kim	100	0.2	0.17
Akamaru T.	100	0.2	
Kim Y.	100	0.2	0.17
Vadapalli S.	12000		
Zander T.	100, 500, 1000, 5000		
Huang H. L.	345.345	0.31	

1.2. FEM Design: Parametrization

The ideal Finite Elemente Model (FEM) of the cervical spine should be versatile and adaptable in an easy manner, and in this way it shouldn't require a high intervention of the programmer in every single case of study. For this purpose, the FEM will be developed using the ANSYS, which allows one to vary the characteristics of the model.

1.2.1. Geometric Parametrization

This is done with 2 objectives:

- To achieve the exact dimensions and characteristics of the spine matched with those obtained from anthropomorphic studies by different authors [18, 19].
- To easily apply the model to different anatomic and physiologic characteristics which vary among patients, such as different lordotic curves.

1.2.2. Mesh Parametrization

This is done to increase the control of the programmer, so that the density and position of the nodes can be chosen, and thus by manipulating them it is possible to incorporate surgical implants directly attached to the nodes of the spine structure. Also, by doing so, there will be no need to generate equations for restriction of degrees of freedom, which slows the calculating time. On the other hand, mesh parametrization also allows for the possibility of removing elements in an easy manner, in order to simulate conditions either clinically or postsurgically, such as post-corpectomy instability, as in this case.

1.2.3. Mechanic Parametrization

This enables the possibility of modifying all the characteristics of the materials. This is very useful in case of certain pathologies such as osteoporosis, disc degeneration; or even the possibility of simulation of different behaviour ways of soft tissue, such as ligaments (calculated from multilinear models, either hyperelastic or viscoelastic).

1.3. Cervical Spine FEM for Corpectomy

The generated software draws the FEM through consecutive steps (Figure 1). To generate the nodes and the elements of the vertebral bodies, solid elements are employed for the cancellous bone, while shell-like elements are used for cortical bone. For modelling of the contacts, a 1 *mm* gap has been used between facet joints by means of node-surface contact elements, with a friction coefficient (between surfaces) of μ=0.85. The capsular ligaments, which bond the facet joints, have also been modelled. Finally, for the modelization of the bone graft, an isotropic material has been used between the vertebral endplates of C5 and C7.

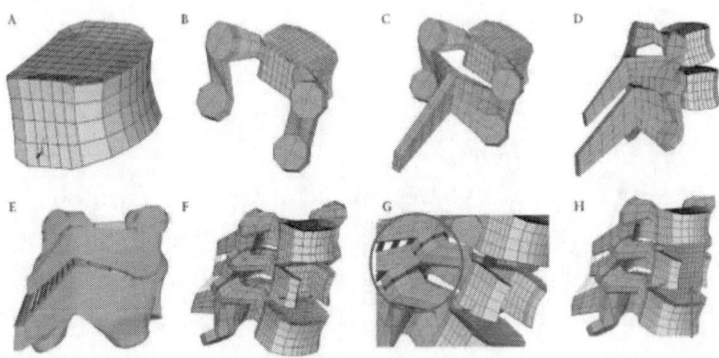

Figure 1. FEM design progression including C6 corporectomy with bone graft C5 -7 (left to right, up to down): generation of (a) nodes and elements of the vertebral body C7; (b) of the pedicles and facets as well as the vertebral body; (c) of laminae and spinous procesess; (d) of the C6 vertebra; (e) of the ligaments: interspinous, supraspinous, anterior and posterior longitudinal as well as the flavum; (f) of the C5 vertebra and C6 corporectomy; (g) of the capsular ligaments; (h) finally, the completed model after graft bone inserted (red arrow).

1.4. FEM of the Osteosynthesis

The procedure followed for the FEM development of the dynamic plate and the screws was different from that of the cervical spine; in order to simulate the sliding movement of the screws over the plate (references: for the ABC plate (Aesculap), ABC – FJ759T of 40 *mm* length; and for the unicortical screws, FJ812T of 12 *mm* length and 4 mm diameter. The modelling process (Figure 2) starts with the geometry of the plate and the head of the screw with the Software program "Solidworks 2008" (Dassault Systèmes). Once obtained, the geometric characteristics are imported into the "ANSYS" program and meshed in a free manner. Finally, the contact surfaces between the head of the screw and the slot are defined. For the dynamic plate slot modelling, contact elements between surface-surface between the head of the screw and the slot have been defined, mimicking the sliding of the screws which occurs during settling or shortening of the bone graft (and allows for the continuous load transmission). After designing and assembling all the elements, the model of cervical fusion with the bone graft filling the gap after corpectomy is obtained (Figure 3).

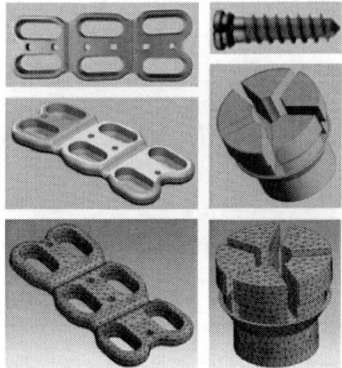

Figure 2. Sequential modelization (up to down) of the instrumentation from an ABC dynamic plate (on the left) and unicortical screw (right) used for the FEM development using Solidworks; this is an intermediate step before transforming them into FEM.

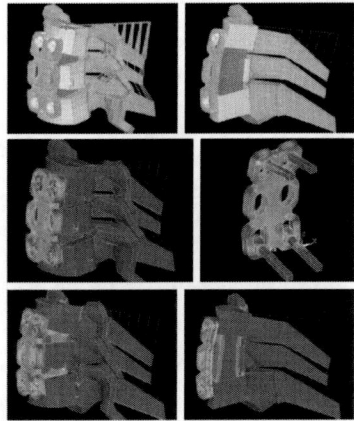

Figure 3. FEM model of anterior cervical corpectomy and fusion with plate screwed to the vertebral bodies (up) in the after immediate postoperative situation, both in static (center) and dynamic (down) configuration.

1.5. Study Design

For each of the two working configurations (static and dynamic), 2 theoretical scenarios were defined, following the biology of graft integration, as after some time the graft length is shortened due to either true shortening because of absorption, or because of settling of the same between the two endplates. In order to be able to compare the results of this study against others already published in the literature, it was decided to shorten the graft 10%. Each one of the scenarios of the FEM (4 in all) has been tested under physiological conditions in flexion applied on C5 through a rigid structure.

2. RESULTS

2.1. FEM Validation

Validation was performed following 3 papers published in the literature [8, 9, 20], using the same conditions. Once the model was validated, these conditions could be modified to simulate any loading condition. The rotational movements given by other authors, as well as geometric and mechanic parameters, were quite variable; so the model was built from a combination of the data taken from different studies. To perform the validation of the model, the mechanical parameters of the ligaments were gradually modified. As a parameter for the validation of the analytical model of the spine, rotation about the axis (applying a pure moment) of a functional vertebral unit was used, with an X axis for flexion and extension, and Z axis for torsion. The cervical spine model was considered validated if the rotational ranges were within the range of motion of the other studies. The loads used for validation were moments between 0,2 Nm and 2 Nm with 0,2 Nm increases, and the obtained values are shown in Table 3; despite large variability of the previously published results, the rotational curve of our model falls inside the range of motions established for validation (Figure 4). Nevertheless, the validation of the

FEM of the plate and screws required the comparison of data compared with those from biomechanical essays.

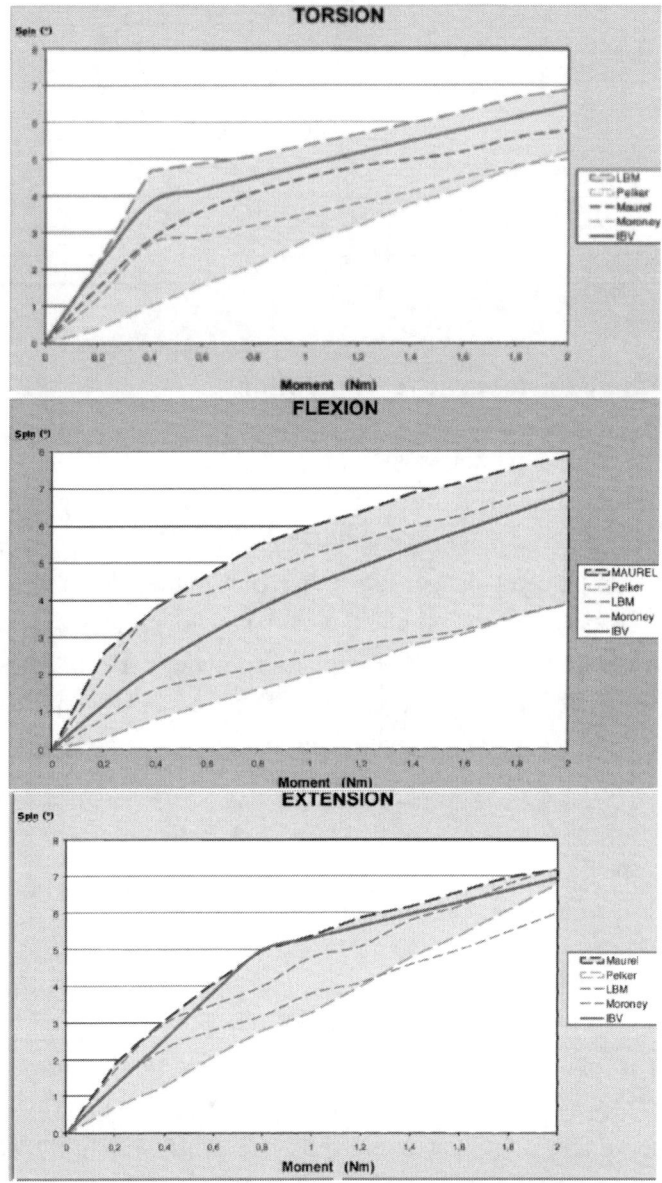

Figure 4. Validation of the FEM under torsion, flexion and extension.

Table 3. Properties of the ligaments of the cervical region C2-4

Moment (Nm)	Flexion spin (°)	Extension spin (°)	Torsion spin (°)
0.2	1.17	1.28	2.06
0.4	2.2	2.56	3.81
0.6	0.04	3.83	4.16
0.8	3.7	4.99	4.5
1	4.38	5.32	4.84
1.2	4.9	5.65	5.17
1.4	5.41	5.97	5.49
1.6	5.89	6.3	5.82
1.8	6.37	6.63	6.14
2	6.85	6.95	6.45

2.2. Immediate Postoperative Simulation

Once the whole model was built, it was tested both in static and dynamic configurations (Figure 3). As the Von Mises equivalent tensions figures show, when the plate is static it is the plate which bears most of the load, with a small load proportion being transmitted down through the graft. The locations of the implant which had more tensions were the contacts between the screws with the plate, and where the thickness of the plate was smaller. When a dynamic plate was used the load was transmitted both through the plate and the graft, with a much larger proportion going through the graft.

2.3. Simulation after Graft Shortening (Subsidence)

After simulation of graft shortening (subsidence) of 10% (Figure 5), most of the load transferred from C5 to C7 was done through the plate when a static plate was used, which can't adapt to the new situation. Alternatively, when the dynamic plate was tested, contact continued between the vertebral endplates and the bone graft, which allowed load-sharing and transfer in the craniocaudal direction, particularly in the anterior part of the graft because of the anatomic flexion movement of the cervical spine. When analysing the Von Mises equivalent tensions (Figure 6) in percentage for the thinnest of the plate, it can be observed how the plate is overloaded with the static plate is used.

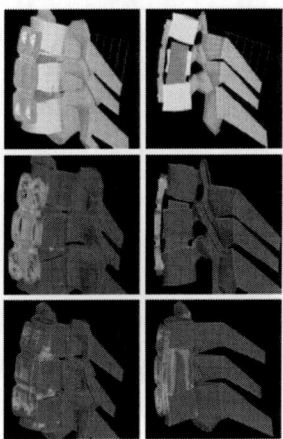

Figure 5. FEM model of anterior cervical corpectomy and fusion with plate screwed to the vertebral bodies (up) after simulated graft shortening, both in static (center) and dynamic (down) configuration.

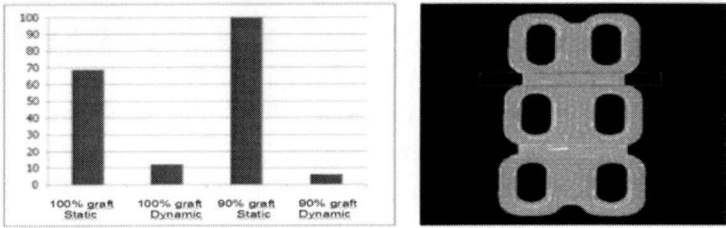

Figure 6. Tension differences (left) on the critical section of the cervical plate (right).

3. DISCUSSION

Biomechanical evaluation of spinal implants can best be achieved with human specimens, but lack of availability of these specimens has boosted the development of other ways of testing. The technique of finite element modelling integrates physical, geometric and mechanical properties through a mathematical model and uses computer simulation to test biomechanical properties. Finite Elements Models (FEM) have proved to be an interesting tool to test spinal implants, and have already been used to assess the cervical spine[21]. Likewise, there are numerous biomechanical studies assessing many aspects of motion and stiffness [18-40]. No studies have been described using or validating Finite Element Modeling to study clinically available dynamic cervical plates. Finite Elements Models (FEM) have proved to be an interesting tool to test spinal implants and have already been used in the cervical spine area. Our study reflects the benefit of dynamic plates in graft loading with flexion; we chose to test our model in flexion-compression because fixed plates tend to stress shield and distract fusion surfaces with this movement as seen in biomechanical and FEM works [41]: paradoxical loads occur when plating, and dynamic plates lessen this effect while keeping uniform load (benefit of plating) seen at the contact pressure plot[42].

Our study reflects the benefit of dynamic plates in graft loading with flexion; we chose to test our model in flexion-compression because fixed plates tend to stress shield and distract fusion surfaces with this movement as seen in biomechanical and FEM works [41]: paradoxical loads occur when plating, and dynamic plates lessen this effect while keeping uniform load (benefit of plating) seen at the contact pressure plot [42].

Several traslational dynamic plates have been shown to have better clinical results, particularly in cases of multilevel fusion with either auto or allograft. This is shown with different dynamic plate styles, bidirectional [43, 44] or unidirectional [45, 46] even when compared with static plates plus autograft [47]. They may also be used for one- or multi- level ACDF with excellent results. Some designs (including the one used in this study) allow for the screws to slide (traslate) and to toggle, even after maximum translation in the slots has been achieved. Clinically, settling occurs in the first weeks or months after surgery, particularly when allograft is used, and averages 2 *mm* per level. Dynamic plates seem to allow faster fusion to occur without hardware failure or loss of kyphosis correction (within 5 degrees), as is sometimes seen with uninstrumented fusion [48]. Even in traumatic cases, with presumed instability, it has been shown that dynamic plates result in less settling and better lordosis preservation with no hardware failures [49]. Some authors even advocate for their use in circumferential fusions [50], with improved results in comparison with static plates [51].

Biomechanical comparisons have been made in the past among different static and dynamic plates, either rotational or uni- or bi-directional traslational [52-58]. Reduction of graft height in prior studies was used to simulate bone resorption or subsidence [59]. The dynamic construct allowed graft contact and improved stability over the static construct with 10% shortening of the graft. Likewise, with the simulated graft shortening (of 10%), the static plates were not able to maintain load sharing [60], resulting in a marked increase in flexion-extension. Our study confirms improved biomechanical performance for dynamic plates, particularly after graft shortening, both in the polyethylene model and in the FEM [14], with decreased stiffness after simulated graft shortening with the plate in static configuration. The main clinical concerns regarding dynamic traslational plates, beyond the scope of this work, and which may be related to plate design are adjacent level impingement due to plate translation during subsidence (which may disappear with modern plate designs) and excessive range of motion prior to healing, which could be disadvantageous in cases of instability. Despite these concerns, this study confirms improved overall stiffness, even after subsidence, with a traslationally dynamic plate, as compared with the same plate placed in static mode.

The use of dynamic cervical plates is based in a theoretical biomechanic superiority which promotes better clinical results. Through this FEM, validated with the others pre-existent cadaver models published in the literature, a cervical corpectomy and fusion model with instrumentation has been configured; and by using proper software which can simulate different scenarios, the situations where dynamic plates could be more appropriate can be simulated.

CONCLUSION

The results obtained in this study after the simulations performed with the plate in either static or dynamic configuration, with full-length of the graft (mimicking immediate postoperative status) or after 10% shortening (simulating graft subsidence), allows us to reach the following conclusions from a biomechanical viewpoint, which can be traslated to the clinical arena:

- In the immediate postoperative timepoint, dynamic plates transfer higher load through the graft than static ones.
- After graft shortening, static plates do not transfer load to the graft, while dynamic plates are able to adapt to the new geometry and allow for load transfer.

REFERENCES

[1] Yoganandan, N., Kumaresan, S., Voo, L. & Pintar, F. A. (1996). Finite element applications in human cervical spine modeling. *Spine*, *21*, 1824-34.

[2] Yoganandan, N., Kumaresan, S. C., Voo, L., Pintar, F. A. & Larson, S. J. (1996). Finite element modeling of the C4-C6 cervical spine unit. *Med Eng Phys*, *18*, 569-74.

[3] Yoganandan, N., Kumaresan, S., Voo, L. & Pintar, F. A. (1997). Finite element model of the human lower cervical spine: parametric analysis of the C4-C6 unit. *J Biomech Eng*, *119*, 87-92.

[4] Kumaresan, S., Yoganandan, N. & Pintar, F. A. (1999). Finite element analysis of the cervical spine: a material property sensitivity study. *Clin Biomech*, *14*, 41-53.

[5] Kumaresan, S., Yoganandan, N., Pintar, F. A. & Maiman, D. J. (1999). Finite element modeling of the cervical spine: role of intervertebral disc under axial and eccentric loads. *Med Eng Phys*, *21*, 689-700.

[6] Stemper, B. D., Yoganandan, N. & Pintar, F. A. (2004). Validation of a head-neck computer model for whiplash simulation. *Med Biol Eng Comput*, *42*, 333-8.

[7] Voo, L., Denman, J., Yoganandan, N., Pintar, F. A. & Cusick, J. F. (1995). A 3-D FE model of the cervical spine with CT-based geometry. *Adv Bioeng*, *29*, 323-24.

[8] Maurel, N. (1993). *Modelisation geometrique et mechanique tridimensionnelle par elements finis du rachis cervical inferieur [Three-dimensional geometric and mechanical finite element modeling of the lower cervical spine]*. L'Ecole Nationale Superieur d'Arts et Metiers. Thesis.

[9] Moroney, S. P., Schultz, A. B., Miller, J. A. & Andersson, G. B. (1988). Load displacement properties of lower cervical spine motion segments. *J Biomech*, *21*, 769-79.

[10] Yoganandan, N., Kumaresan, S. & Pintar, F. A. (2000). Geometric and mechanical properties of human cervical spine ligaments. *J Biomech Eng*, *122* (6), 623-9.

[11] Yoganandan, N., Kumaresan, S. & Pintar, F. A. (2001). Biomechanics of the cervical spine Part 2. Cervical spine soft tissue responses and biomechanical modeling. *Clin Biomech*, *16*, 1-27.

[12] Kim, Y. (2007). Finite element analysis of anterior lumbar interbody fusion: threaded cylindrical cage and pedicle screw fixation. *Spine*, *32*, 2558-68.

[13] Akamaru. T., Kawahara, N., Sakamoto, J., Yoshida, A., Murakami, H., Hato, T., Awamori, S., Oda, J. & Tomita, K. (2005). The transmission of stress to grafted bone inside a titanium mesh cage used in anterior column reconstruction after total spondylectomy: a finite-element analysis. *Spine*, *30*, 2783-7.

[14] Kim, Y. (2001). Prediction of mechanical behaviors at interfaces between bone and two interbody cages of lumbar spine segments. *Spine*, *26*, 1437-42.

[15] Vadapalli, S., Sairyo, K., Goel, V. K., Robon, M., Biyani, A., Khandha, A. & Ebraheim, N. A. (2006). Biomechanical rationale for

using polyetheretherketone (PEEK) spacers for lumbar interbody fusion-A finite element study. *Spine, 31*, E992-8.

[16] Zander, T., Rohlmann, A., Klöckner, C. & Bergmann, G. (2002). Effect of bone graft characteristics on the mechanical behavior of the lumbar spine. *J Biomech, 35*, 491-7.

[17] Huang, H. L., Fuh, L. J., Hsu, J. T., Tu, M. G., Shen, Y. W. & Wu, C. L. (2008). Effects of implant surface roughness and stiffness of grafted bone on an immediately loaded maxillary implant: a 3D numerical analysis. *J Oral Rehabil, 35*, 283-90.

[18] Panjabi, M. M., Duranceau, J., Goel, V., Oxland, T. & Takata, K. (1991). Cervical human vertebrae. Quantitative three-dimensional anatomy of the middle and lower regions. *Spine, 16*, 861-9.

[19] Tan, S. H., Teo, E. C. & Chua, H. C. (2004). Quantitative three-dimensional anatomy of cervical, thoracic and lumbar vertebrae of Chinese Singaporeans. *Eur Spine J, 13*, 137-46.

[20] Pelker, R. R., Duranceau, J. S. & Panjabi, M. M. (1991). Cervical spine stabilization. A three-dimensional, biomechanical evaluation of rotational stability, strength, and failure mechanisms. *Spine, 16*, 117-22.

[21] Pitzen, T. R., Matthis, D., Barbier, D. D. & Steudel, W. I. (2002). Initial stability of cervical spine fixation: predictive value of a finite element model. *Journal of Neurosurgery, 97*(S1), 128-34.

[22] Zhang, Q. H., Teo, E. C., Ng, H. W. & Lee, V. S. (2006). Finite element analysis of moment-rotation relationships for human cervical spine. *Journal of Biomechanics, 39*(1), 189-93.

[23] Ha, S. K. (2006). Element modeling of multi-level cervical spinal segments (C3-C6) and biomechanical analysis of an elastomer-type prosthetic disc. *Medical Engineering Med Eng Phys, 28*(6), 534-41.

[24] Gilbertson, L. G., Goel, V. K., Kong, W. Z. & Clausen, J. D. (1995). Finite element methods in spine biomechanics research. *Crit Rev Biomed Eng, 23*(5-6), 411-73.

[25] Clausen, J. D., Goel, V. K., Traynelis, V. C. & Scifert, J. (1997). Uncinate processes and Luschka joints influence the biomechanics of

the cervical spine: quantification using a finite element model of the C5-C6 segment. *J Orthop Res*, *15*(3), 342-7.

[26] Goel, V. K. & Clause, J. D. (1998). Prediction of load sharing among spinal components of a C5-C6 motion segment using the finite element approach. *Spine*, *23*(6), 684-91.

[27] Faizan, A., Goel, V. K., Biyani, A., Garfin, S. R. & Bono, C. M. (2012). Adjacent level effects of bi level disc replacement, bilevel fusion and disc replacement plus fusion in cervical spine: A finite element based study. *Clin Biomech*, *27*(3), 226-33.

[28] Faizan, A., Goel, V. K., Garfin, S. R., Bono, C. M., Serhan, H., Biyani, A., Elgafy, H., Krishna, M. & Friesem, T. (2012). Do design variations in the artificial disc influence cervical spine biomechanics? A finite element investigation. *Eur Spine J*, *21*, S653-62.

[29] Maurel, N., Lavaste, F. & Skalli, W. (2005). A three-dimensional parameterized finite element model of the lower cervical spine. Study of the influence of the posterior articular facets. *J Biomech*, *38*(9), 1865-72.

[30] Frechede, B., Bertholon, N., Saillant, G., Lavaste, F. & Skalli, W. (2006). Finite element model of the human neck during omnidirectional impacts. Part II: relation between cervical curvature and risk of injury. *Computer Methods in Biomechanical and Biomedical Engineering*, *9*(6), 379-86.

[31] Rousseau, M. A., Bonnet, X. & Skalli, W. (2008). Influence of the geometry of a ball-and-socket intervertebral prosthesis at the cervical spine: a finite element study. *Spine*, *33*(1), E10-4.

[32] Laville, A., Laporte, S. & Skalli, W. (2009). Parametric and subject-specific finite element modelling of the lower cervical spine. Influence of geometrical parameters on the motion patterns. *J Biomech*, *22*, 42(10), 1409-15.

[33] Yoganandan, N., Kumaresan, S. C., Voo, L., Pintar, F. A. & Larson, S. J. (1996). Finite element modelling of the C4-C6 cervical spine unit. *Medical Engineering Physics*, *18*(7), 569-74.

[34] Yoganandan, N., Kumaresan, S. & Voo, L. (1996). Finite Element Applications in Human Cervical Spine Modelling. *Spine*, *21*(15), 1824-1834.

[35] Kumarasen, S., Yoganandan, N. & Pintar, F. A. Finite element analysis of anterior cervical spine interbody fusion. *Biomedical Materials Engineering*, *7*(4), 221-30.

[36] Kumarasen, S., Yoganandan, N., Pintar, F. A., Voo, L. M., Cusick, J. F. & Larson, S. J. (1997). Finite element modelling of cervical laminectomy with graded facetectomy. *Journal of Spinal Disorders*, *10*(1), 40-6.

[37] Kumaresan, S., Yoganandan, N. & Pintar, F. A. (1998). Finite element modeling approaches of human cervical spine facet joint capsule. *J Biomech*, *31*(4), 371-6.

[38] Kumarasen, S., Yoganandan, N., Pintar, F. A., Maiman, D. J. & Kuppa, S. (2000). Biomechanical study of paediatric human cervical spine, a finite element approach. *Journal of Biomechanical Engineering*, *122*(1), 60-71.

[39] Wheeldon, J., Khouphongsy, P., Kumaresan, S., Yoganandan, N. & Pintar, F. A. (2000). Finite element model of human cervical spinal column. *Biomed Sci Instrum*, *36*, 337-42.

[40] Wheeldon, J. A., Stemper, B. D., Yoganandan, N. & Pintar, F. A. (2008). Validation of a finite element model of the young normal lower cervical spine. *Ann Biomed Eng*, *36*(9), 1458-69.

[41] DiAngelo, D. J., Foley, K. T., Vossel, K. A., Rampersaud, Y. R. & Jansen, T. H. (2000). Anterior Cervical Plating Reverses Load Transfer through Multilevel Strut-Grafts. *Spine*, *25*, 783–795.

[42] Galbusera, F., Bellini, C. M., Costa, F., Assietti, R. & Fornari, M. (2008). Anterior cervical fusion: a biomechanical comparison of 4 techniques. Laboratory investigation. *J Neurosurg Spine*, *9*(5), 444-9.

[43] Bose, B. (2003). Anterior cervical arthrodesis using DOC dynamic stabilization implant. *Journal of Neurosurgery* (Spine 1), *98*, 8–13.

[44] Ghahreman, A., Rao, P. J. V. & Ferch, R. D. (2009). Dynamic Plates in Anterior Cervical Fusion Surgery, Graft Settling and Cervical Alignment. *Spine*, Volume *34*, 15, 1567–1571.

[45] Nunley, P. D., Jawahar, A., Kerr, E. J., Cavanaugh, D. A., Howard, C. & Brandao, S. M. (2009). Choice of plate may affect outcomes for single versus multilevel ACDF: results of a prospective randomized single-blind trial. *The Spine Journal*, *9*, 121–127.

[46] Kuklo, T., Rosner, M. & Neal, C. (2005). Two-year sagittal cervical evaluation of static versus dynamic anterior cervical plates. *Spine*, *7*, 191–192.

[47] Goldberg, G., Albert, T. J., Vaccaro, A. R., Hilibrand, A. S., Anderson, D. G. & Wharton, N. (2007). Short-term Comparison of Cervical Fusion With Static and Dynamic Plating Using Computerized Motion Analysis. *Spine*, *32*, 13, 371–5.

[48] Steinmetz, M. P., Benzel, E. C. & Apfelbaum, R. (2005), Subsidence and dynamic cervical spine stabilization. In: *Spine Surgery: techniques, complication avoidance and management*. 2nd ed. Elsevier, Philadelphia.

[49] Khoo, L. T., Benae, J. L. & Gravori, T. (2005). Anterior Plating for Cervical Traumatic Fractures: An Analysis of Graft Height and Segmental Lordosis Preservation. *The Internet Journal of Spine Surgery*, *1*, 2.

[50] Aryan, H. E., Sanchez-Mejia, R. O., Ben-Haim, S. & Ames, C. P. (2007). Successful treatment of cervical myelopathy with minimal morbidity by circumferential decompression and fusion. *Eur Spine J*, *16*(9), 1401-9.

[51] Epstein, N. E. (2003). Fixed vs dynamic plate complications following multilevel anterior cervical corpectomy and fusion with posterior stabilization. *Spinal Cord*, *41*(7), 379-84.

[52] Rapoff, A. J., O'Brein, T. J., Ghanayem, A. J., Heisey, D. M. & Zdeblick, T. A. (1999). Anterior cervical graft and plate load sharing. *J Spinal Disord*, *12*, 45-9.

[53] Saphier, P. S., Arginteanu, M. S., Moore, F. M., Steinberger, A. A. & Camins, M. B. (2007). Stress-shielding compared with

loadsharing anterior cervical plate fixation: a clinical and radiographic prospective analysis of 50 patients. *Journal of Neurosurgery (Spine)*, 6, 391–397.

[54] Truumees, E., Demetropoulos, C. K., Yang, K. H. & Herkowitz, H. N. (2003). Effects of a Cervical Compression Plate on Graft Forces in an Anterior Cervical Discectomy Model. *Spine*, 28(11), 1097–1102.

[55] Cheng, B. C., Burns, P., Pirris, S. & Welch, W. C. (2009). Load sharing and stabilization effects of anterior cervical devices. *J Spinal Disord Tech*, 22(8), 571-7.

[56] Dvorak, M. F., Pitzen, T., Zhu, Q., Gordon, J. D., Fisher, C. G. & Oxland, T. R. (2005). Anterior cervical plate fixation: a biomechanical study to evaluate the effects of plate design, endplate preparation, and bone mineral density. *Spine*, 30(3), 294-301.

[57] Welch, W. C. (2009). Load sharing and stabilization effects of anterior cervical devices. *J Spinal Disord Tech.*, 22(8), 571-7.

[58] Kirkpatrick, J. S., Levy, J. A., Carillo, J. & Moeini, S. R. (1999). Reconstruction after multilevel corpectomy in the cervical spine. A sagittal plane biomechanical study. *Spine*, 24(12), 1186-90; discussion 1191.

[59] Reidy, D., Finkelstein, J., Nagpurkar, A., Mousavi, P. & Whyne, C. (2004), Cervical spine loading characteristics in a cadaveric C5 corporectomy model using a static and dynamic plate. *J Spinal Disord Tech*, 17(2), 117-22.

[60] Brodke, D. S., Gollogly, S., Alexander, R., Nguyen, B. K., Dailey, A. T. & Bachus, A. K. (2001). Dynamic cervical plates: biomechanical evaluation of load sharing and stiffness. *Spine*, 15, 26 (12), 1324-9.

In: A Closer Look at Biomechanics
Editor: Daniela Furst

ISBN: 978-1-53615-866-3
© 2019 Nova Science Publishers, Inc.

Chapter 3

BIOMECHANICAL BEHAVIOUR OF TRASLATIONAL DYNAMIC CERVICAL PLATES. PART 2: BIOMECHANICAL ESSAYS

Javier M. Duart[1],*, MD, Werner Schmoelz[2], PhD, Carlos Atienza[3], EngD, Ignacio Bermejo[3], EngD, Darrel S. Brodke[4], MD, Tobias Pitzen[5], MD and Julio V. Duart[6], MD

[1]Spine and Neurosurgical Services, General Hospital, Valencia, Spain. Departament de Cirugia, UAB, Spain
[2]Dept. of Trauma Surgery, Medical University of Innsbruck, Innsbruck, Austria
[3]Institute of Biomechanics of Valencia, Valencia, Spain.
[4]Orthopedic Department, University of Utah, US
[5]SRH Klinikum Karlsbad-Langensteinbach, Karlsbad, Germany
[6]Hospital de Estella, Estella. Navarra, Spain

---

* Corresponding Author's E-mail: jduart@uv.es.

ABSTRACT

Fusion following anterior cervical decompressive procedures (such as discectomy and corpectomy) frequently involves bone and osteosynthesis with a plate in the case of corpectomies, where either a structural graft (from iliac crest or fibula) or a containing bone cylinder mesh are secured with an anterior plate to avoid extrusion and increase fusion chance. Plate design has evolved from unconstrained static to dynamic or semiconstrained plates, which in turn can be either rotational or traslational, with uni or bidirectional dynamicity. Apart from the clinical studies in the literature supporting dynamic plates, some biomechanical studies also favour them in comparison with static plates regarding load transmission which is thought to foster graft integration; nevertheless, hardly any study addresses the same plate design working under static and dynamic configuration, in order to discard material or geometrical properties which could explain part of the superior results both in vivo and in vitro scenarios.

This study was arranged, to asess the biomechanical behaviour with mechanical essays. The purpose of this research work was to simulate static and dynamic behavior with the same anterior cervical plate design and in two different clinical scenarios, both in the immediate postoperative and after simulated graft subsidence (four simulated clinical scenarios in all) by means of biomechanical essays, mimicking in vivo situations. Results show that load transmission is superior when the plate works dynamically, particularly after shortening of the graft.. so dynamic plates confer biomechanical advantages by improving transfer load and adapting to graft shortening.

Keywords: biomechanics, corpectomy, dynamic plate

INTRODUCTION

Anterior cervical decompressive procedures, such as discectomy (ACDF) and corpectomy (ACF) are usually followed by fusion with bone graft (or substitutes) and then secured with an anterior plate to avoid extrusion and increase the fusion rate (plating is advised with corpectomy and an option with cage). Plate fixation has been shown to improve fusion rates and clinical results particularly in multilevel procedures [1, 2], while in one-level procedures its use is more debatable [3, 4], as it keeps sagittal alignment [5], preventing subsidence [6, 7] and decreasing reoperation rate due to fewer graft-related complications. Cervical plate design has evolved significantly over the last 2 decades [8] to manage hardware associated complications [9-13]. Initial static, unconstrained plates had risks of screw breakage and pull-out, and required bicortical fixation. Constrained plates, with locked screws, evolved to prevent hardware loosening but in some cases lead to stress-shielding which may decrease fusion rate due to by-passing of the forces on the graft, required to promote fusion, according to Wolff's Law. Finally, the semi-constrained "dynamic" plates evolved, allowing for loading-sharing, thus promoting fusion by fostering graft integration.

These dynamic plates can be either rotational or traslational, with uni- or bi-directional movement between the screws and plate, and are thought to have the best clinical results with shorter fusion times, higher fusion rates, and fewer hardware complications by allowing higher graft loading even after subsidence, which frequently occurs as part of the healing process of integration of the graft. While several cervical plate systems have been designed to provide the surgeon choice between static and dynamic constructs, to our knowledge no *in vitro* cadaveric study has tested the biomechanical behavior of a bidirectional traslational plate against the exact same plate in static configuration. Our group has published a finite element model (FEM) to explore this topic previously [14-16]. The purpose of this study is to investigate the *in vitro* biomechanical differences between a traslational dynamic plate and the same plate in static construct, using cadaveric biomechanical assessment and its FEM for validation.

METHODS

A C6 corpectomy was simulated with plastic blocks: following ISO standards, an interbody graft of UHMWPE was placed between the simulated C5 and C7 vertebrae (Figure 1) and the construct was mounted on our materials testing machine.

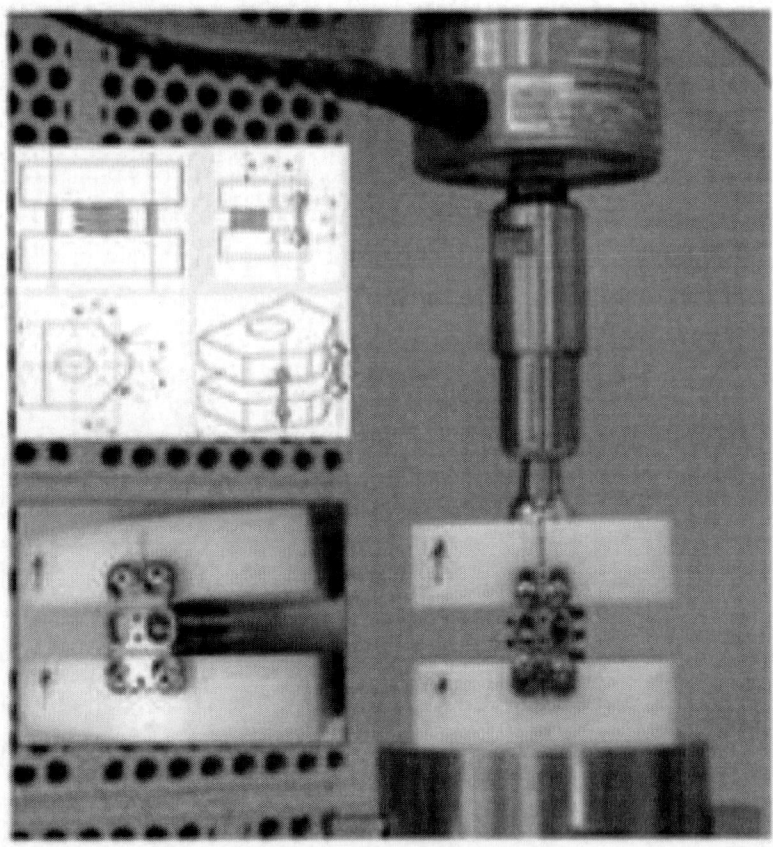

Figure 1. Biomechanical essays with polyethylene blocks; left up, diagram of blocks; left down, dynamic configuration; right, INSTROM machine with the plate in static configuration (note the bolts blocking dynamism).

An ABC™ slotted dynamic cervical plate (Aesculap - Tuttlingen, Germany) was then placed. Four different clinical scenarios were tested - two different graft lengths (full length and simulated 10% subsidence) with the plate placed in either dynamic mode (natural) or static mode (locked), by blocking the movement of the screws in the slots. Stiffness in flexion-compression and displacement between plate and vertebral body were evaluated following the ISO12189: 2008 rules. For each configuration, three repetitions were done in displacement control with a speed of 0.42mm/s, ending after plate failure; with the obtained data, a load-displacement curve was drawn, and data regarding displacement (mm) and load (N) to the elastic limit with flexion-compression, as well as stiffness (N/mm) and load to failure. The second objective was validation of the FEM.

For the plate, values obtained in the mechanical tests were compared with those obtained analytically from the models of the dynamic plate, screws and polyethylene blocks developed and used under the same ISO load and environment conditions (Figure 2).

Figure 2. Up and middle, polyethylene blocks and whole construct with implanted cervical plate developed firstly with Ansys and then the FEM, left and right respectively. Below, simulated displacement of the construct in the direction of the load application.

The same loads using a zenithal sphere were applied in the validation process of the FEM, for each of the 4 configurations (adjustments have been made for static and dynamic configurations). The models were drawn using CAD Software for the same blocks used in the essays and the assembled plate. They were then imported using ANSYS Workbench software and meshing was performed, as well as the different contact points.

After this was accomplished, the same loads were applied as in the mechanical essay; the superior hemisphere on which load was applied could rotate freely in the block (friction coefficient zero), and the union between the screws and the plate was totally rigid. The inferior surface of the block was considered fixed, and vertical load was applied on the hemisphere. For the adjustment of the FEM, data were taken from the experimental data of the mechanical essays of the plate in one of the configurations (static in full length); rigidity of the elastic assembly was obtained from these essays, and the model was adjusted by making variations in the rigidity of the screws and the spring. Once validation had been done, tensions were analysed, regarding displacement in the direction of the applied load (with a maximal displacement of 0.019mm).

RESULTS

Force-Displacement data obtained reveal that when the plate is in dynamic configuration (n. 1 & 2), it allows the same displacement at a lower force; when the graft is shortened (simulated subsidence), the dynamic plate is able to allow for continuing graft loading (simulated with a spring), as opposed to the static plate (n. 3 & 4) which requires a longer distance until contact is provided since the plate is not able to adapt and the stiffness measured corresponds only to the plate (Figure 3). Mean stiffness values obtained after 3 repetitions for each configuration (and standard deviations) were as follows (N/mm): 89.48 (+ 2.65) and 94.77 (+ 1.16) for dynamic and 95.47 (+ 4.7) and 84.13 (+ 8.1) for static configuration, respectively; the only statistical difference was found for configurations testing the plate in static configuration (p = 0.03395).

Validation was done for the model using FEM under the four scenarios with the computer work, showing no difference with the biomechanical testing.

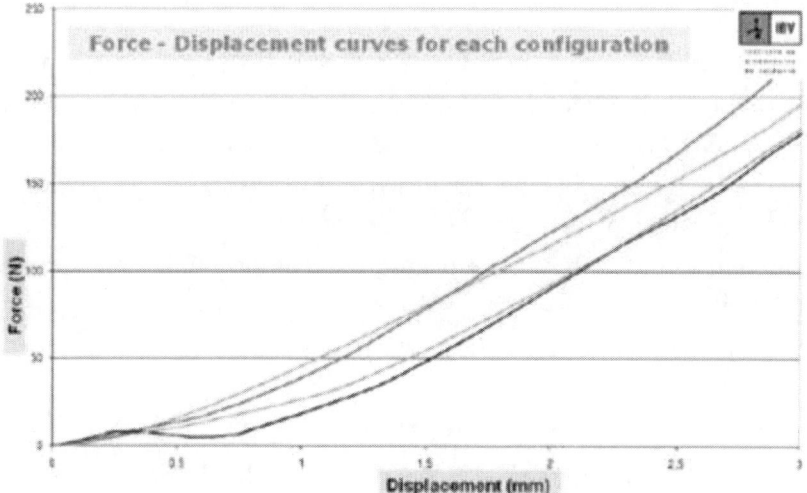

Figure 3. Force-displacement curves for each configuration. Red and orange lines refer to static configuration with full and shortened lengths. Green is for dynamic configuration with full length, showing that dynamic plates allow higher load transmission in the immediate postoperative period; blue is for dynamic configuration with shortened length, showing that these dynamic plates are able to accommodate to the new geometry after graft shortening and keep on with load transmission through it.

DISCUSSION

Biomechanical evaluation of spinal implants can best be achieved with human specimens, but lack of availability of these specimens has boosted the development of other ways of testing.

The technique of finite element modelling integrates physical, geometric and mechanical properties through a mathematical model and uses computer simulation to test biomechanical properties. Finite Elements

Models (FEM) have proved to be an interesting tool to test spinal implants, and have already been used to assess the cervical spine [17]. Likewise, there are numerous biomechanical studies assessing many aspects of motion and stiffness [18-40]. No studies have been described using or validating Finite Element Modeling to study clinically available dynamic cervical plates. Finite Elements Models (FEM) have proved to be an interesting tool to test spinal implants, and have already been used in the cervical spine area. Our study reflects the benefit of dynamic plates in graft loading with flexion; we chose to test our model in flexion-compression because fixed plates tend to stress shield and distract fusion surfaces with this movement as seen in biomechanical and FEM works [41]: paradoxical loads occur when plating, and dynamic plates lessen this effect while keeping uniform load (benefit of plating) seen at the contact pressure plot [42].

Several traslational dynamic plates have been shown to have better clinical results, particularly in cases of multilevel fusion with either auto or allograft. This is shown with different dynamic plate styles, bidirectional [43, 44] or unidirectional [45, 46] even when compared with static plates plus autograft [47]. They may also be used for one- or multi- level ACDF with excellent results. Some designs (including the one used in this study) allow for the screws to slide (traslate) and to toggle, even after maximum translation in the slots has been achieved. Clinically, settling occurs in the first weeks or months after surgery, particularly when allograft is used, and averages 2 mm per level.

Dynamic plates seem to allow faster fusion to occur without hardware failure or loss of kyphosis correction (within 5 degrees), as is sometimes seen with uninstrumented fusion [48]. Even in traumatic cases, with presumed instability, it has been shown that dynamic plates result in less settling and better lordosis preservation with no hardware failures [49]. Some authors even advocate for their use in circumferential fusions [50], with improved results in comparison with static plates [51].

Biomechanical comparisons have been made in the past among different static and dynamic plates, either rotational and uni- or bi-directional traslational [52-58]. Reduction of graft height in prior studies was used to simulate bone resorption or subsidence [59]. The dynamic construct allowed graft contact and improved stability over the static construct with 10% shortening of the graft. Likewise, with the simulated graft shortening (of 10%), the static plates were not able to maintain load sharing [60], resulting in a marked increase in flexion-extension. Our study confirms improved biomechanical performance for dynamic plates, particularly after graft shortening, both in the polyethylene model and in the FEM, with decreased stiffness after simulated graft shortening with the plate in static configuration.

The main clinical concerns regarding dynamic traslational plates, beyond the scope of this work, and which may be related to plate design are adjacent level impingement due to plate translation during subsidence (which may disappear with modern plate designs) and excessive range of motion prior to healing, which could be disadvantageous in cases of instability. Despite them, this study confirms improved overall stiffness, even after subsidence, with a traslationally dynamic plate, as compared with the same plate placed in static mode.

CONCLUSION

From the data obtained in this biomechanical study, we can conclude that:

- Traslational dynamic plates allow higher load transmission in the immediate postoperative period, and after shortening (subsidence) of the graft they can accommodate to the new geometry and keep load transmission through the graft. This cannot occur with a static plate.
- There are higher stresses in the static plates, which increase over time after the operation as the graft shortens.

ACKNOWLEDGMENT

This work has been possible thanks to a research grant of Trauma Foundation.

REFERENCES

[1] Wang, J. C., McDonough, P. W., Endow, K. K., Delamarter, R. B. Increased fusion rates with cervical plating for two-level anterior cervical discectomy and fusion. *Spine,* 2000; 25(1):41 - 5.

[2] Wang, J. C., McDonough, P. W., Kanim, L. E., Endow, K. K., Delamarter, R. B. Increased fusion rates with cervical plating for three-level anterior cervical discectomy and fusion. *Spine,* 2001; 26(6):643 - 6; discussion 646 - 7.

[3] Wang, J. C., McDonough, P. W., Endow, K., Kanim, L. E., Delamarter, R. B. The effect of cervical plating on single-level anterior cervical discectomy and fusion. *J. Spinal Disord.,* 1999; 12(6):467 - 71.

[4] Samartzis, D., Shen, F. H., Lyon, C., Phillips, M., Goldberg, E. J., An, H. S. Does rigid instrumentation increase the fusion rate in one-level anterior cervical discectomy and fusion? *Spine J.,* 2004; 4(6):636 - 43.

[5] Song, K. J., Taghavi, C. E., Lee, K. B., Song, J. H., Eun, J. P. The efficacy of plate construct augmentation versus cage alone in anterior cervical fusion. *Spine,* 2009; 34(26):2886 - 92.

[6] Xie, J. C., Hurlbert, R. J. Discectomy versus discectomy with fusion versus discectomy with fusion and instrumentation: a prospective randomized study. *Neurosurgery,* 2007; 61(1):107 - 1.

[7] Tye, G. W., Graham, R. S., Broaddus, W. C., Young, H. F. Graft subsidence after instrument-assisted anterior cervical fusion. *J. Neurosurg.,* 2002; 97(2 Suppl.):186 - 92.

[8] Haid, R. W., Foley, K. T., Rodts, G. E., Barnes, B. The Cervical Spine Study Group anterior cervical plate nomenclature. *Neurosurg. Focus,* 12 (1):Article 15, 2002.

[9] Fassett, D. R., Csaszar, D. J., Albert, T. J. Anterior cervical plating update. *Current Opinion in Orthopaedics*, 2007; 18:282 - 288.

[10] Lowery, G. L., McDonough, R. F. The Significance of Hardware Failure in Anterior Cervical Plate Fixation: Patients with 2- to 7-Year Follow-up. *Spine,* 1998; 23(2):181 - 186.

[11] Gonugunta, V., Krishnaney, A. A., Benzel, E. C. Anterior cervical plating. *Neurology India,* 2005; 53(4):424 - 32.

[12] Kwon, B. K., Vaccaro, A. R., Grauer, J. N., Beiner, J. M. The use of rigid internal fixation in the surgical management of cervical spondylosis. *Neurosurgery,* 2007; 60(1 Suppl. 1):S118-29.

[13] Rhee, J. M., Park, J. B., Yang, J. Y., Riew, K. D. Indications and techniques for anterior cervical plating. *Neurology India*, 2005; 53(4), 433 - 9.

[14] Duart Clemente, J. M., Atienza Vicente, C. M., Bermejo Bosch, I., Morales Martín, I., Gil Guerrero, I., Duart Clemente, J. V. Simulación mediante modelos de elementos finitos del comportamiento biomecánico de las placas cervicales dinámicas [Simulation using finite element models of the biomechanical behavior of the dynamic cervical plates]. *Trauma Fundación Mapfre*, 2009; 20:4 221 - 228.

[15] Duart, Javier M., Carlos Atienza, Ignacio Bermejo Bosch, Íñigo Morales Martín, Inés Gil Guerrero, Julio V. Duart Clemente. Biomechanical behaviour of traslational dynamic plates: An *in vitro* study. *Global Spine Congress*, 2011.

[16] Duart Clemente, J. M. et al. Superiority of traslational dynamic plates through combined finite element model and biomechical essays. *Spineweek*, 2012.

[17] Pitzen, T. R., Matthis, D., Barbier, D. D., Steudel, W. I. Initial stability of cervical spine fixation: predictive value of a finite element model. *Journal of Neurosurgery*, 2002; 97(S1):128 - 34.

[18] Zhang, Q. H., Teo, E. C., Ng, H. W., Lee, V. S. Finite element analysis of moment-rotation relationships for human cervical spine. *Journal of Biomechanics*, 2006; 39(1):189 - 93.

[19] Ha, S. K. Element modeling of multi-level cervical spinal segments (C3-C6) and biomechanical analysis of an elastomer-type prosthetic disc. *Medical Engineering Physics*, 2006; 28(6):534 - 41.

[20] Gilbertson, L. G., Goel, V. K., Kong, W. Z., Clausen, J. D. Finite element methods in spine biomechanics research. *Crit. Rev. Biomed. Eng.*, 1995; 23(5-6):411 - 73.

[21] Clausen, J. D., Goel, V. K., Traynelis, V. C., Scifert, J. Uncinate processes and Luschka joints influence the biomechanics of the cervical spine: quantification using a finite element model of the C5-C6 segment. *J. Orthop. Res.*, 1997 May;15(3):342 - 7.

[22] Goel, V. K., Clause, J. D. Prediction of load sharing among spinal components of a C5-C6 motion segment using the finite element approach. *Spine,* 1998; 23(6):684 - 91.

[23] Faizan, A., Goel, V. K., Biyani, A., Garfin, S. R., Bono, C. M. Adjacent level effects of bi level disc replacement, bi level fusion and disc replacement plus fusion in cervical spine--a finite element based study. *Clin. Biomech.*, 2012 Mar.; 27(3):226 - 33.

[24] Faizan, A., Goel, V. K., Garfin, S. R., Bono, C. M., Serhan, H., Biyani, A., Elgafy, H., Krishna, M., Friesem, T. Do design variations in the artificial disc influence cervical spine biomechanics? A finite element investigation. *Eur. Spine J.*, 2012 Jun.; 21 Suppl. 5:S653-62.

[25] Maurel, N., Lavaste, F., Skalli, W. A three-dimensional parameterized finite element model of the lower cervical spine. Study of the influence of the posterior articular facets. *J. Biomech.*, 2005; 38(9):1865 - 72.

[26] Frechede, B., Bertholon, N., Saillant, G., Lavaste, F., Skalli, W. Finite element model of the human neck during omnidirectional impacts. Part II: relation between cervical curvature and risk of injury. *Computer Methods in Biomechanical and Biomedical Engineering,* 2006; 9(6):379 - 86.

[27] Rousseau, M. A., Bonnet, X., Skalli, W. Influence of the geometry of a ball-and-socket intervertebral prosthesis at the cervical spine: a finite element study. *Spine,* 2008; 33(1):E10-4.

[28] Laville, A., Laporte, S., Skalli, W. Parametric and subject-specific finite element modelling of the lower cervical spine. Influence of geometrical parameters on the motion patterns. *J. Biomech.*, 2009 Jul. 22; 42(10):1409 - 15.

[29] Yoganandan, N., Kumaresan, S., Voo, L., Pintar, F. A. Finite element applications in human cervical spine modeling. *Spine,* 1996; 21(15):1824 - 34.

[30] Yoganandan, N., Kumaresan, S. C., Voo, L., Pintar, F. A., Larson, S. J. Finite element modelling of the C4-C6 cervical spine unit. *Medical Engineering Physics,* 1996; 18(7):569 - 74.

[31] Yoganandan, N., Kumaresan, S., Voo, L. Finite Element Applications in Human Cervical Spine Modelling. *Spine,* 1996; 21(15):1824 - 1834.

[32] Yoganandan, N., Kumaresan, S., Voo, L., Pintar, F. A. Finite element model of the human lower cervical spine: parametric analysis of the C4-C6 unit. *J. Biomech. Eng.,* 1997; 119(1):87 - 92.

[33] Kumarasen, S., Yoganandan, N., Pintar, F. A. Finite element analysis of anterior cervical spine interbody fusion. *Biomedical Materials Engineering,* 1997; 7(4):221 - 30.

[34] Kumarasen, S., Yoganandan, N., Pintar, F. A. et al. Finite element modelling of cervical laminectomy with graded facetectomy. *Journal of Spinal Disorders,* 1997; 10(1):40 - 6.

[35] Kumaresan, S., Yoganandan, N., Pintar, F. A. Finite element modeling approaches of human cervical spine facet joint capsule. *J. Biomech.,* 1998; 31(4):371 - 6.

[36] Kumaresan, S., Yoganandan, N., Pintar, F. A. Finite element analysis of the cervical spine: a material property sensitivity study. *Clin. Biomech.,* 1999; 14(1):41 - 53.

[37] Kumaresan, S., Yoganandan, N., Pintar, F. A., Maiman, D. J. Finite element modeling of the cervical spine: role of intervertebral disc

under axial and eccentric loads. *Med. Eng. Phys.,* 1999; 21(10):689 - 700.

[38] Kumarasen, S., Yoganandan, N., Pintar, F. A. et al. Biomechanical study of paediatric human cervical spine, a finite element approach. *Journal of Biomechanical Engineering,* 2000; 122(1):60 - 71.

[39] Wheeldon, J., Khouphongsy, P., Kumaresan, S., Yoganandan, N., Pintar, F. A. Finite element model of human cervical spinal column. *Biomed. Sci. Instrum.,* 2000; 36:337 - 42.

[40] Wheeldon, J. A., Stemper, B. D., Yoganandan, N., Pintar, F. A. Validation of a finite element model of the young normal lower cervical spine. *Ann. Biomed. Eng.,* 2008; 36(9):1458 - 69.

[41] DiAngelo, D. J., Foley, K. T., Vossel, K. A., Rampersaud, Y. R., Jansen, T. H. Anterior Cervical Plating Reverses Load Transfer Through Multilevel Strut-Grafts. *Spine,* 2000; 25:783 - 795.

[42] Galbusera, F., Bellini, C. M., Costa, F., Assietti, R., Fornari, M. Anterior cervical fusion: a biomechanical comparison of 4 element approach. *Journal of Biomechanical Engineering,* 2000; 122(1): 60 - 71.

[43] Bose, B. Anterior cervical arthrodesis using DOC dynamic stabilization implant. *Journal of Neurosurgery,* (Spine 1) 2003; 98: 8 - 13.

[44] Ghahreman, A., Rao, P. J. V., Ferch, R. D. Dynamic Plates in Anterior Cervical Fusion Surgery, Graft Settling and Cervical Alignment. *Spine,* Volume 34, Number 15, pp 1567 - 1571, 2009.

[45] Nunley, P. D., Jawahar, A., Kerr, E. J., Cavanaugh, D. A., Howard, C., Brandao, S. M. Choice of plate may affect outcomes for single versus multilevel ACDF: results of a prospective randomized single-blind trial. *The Spine Journal,* 9(2009) 121 - 127.

[46] Kuklo, T., Rosner, M., Neal, C. Two-year sagittal cervical evaluation of static versus dynamic anterior cervical plates. *Spine: Affiliated Society Meeting Abstracts,* 2005: 7, pp 191 - 192.

[47] Goldberg, G., Albert, T. J., Vaccaro, A. R., Hilibrand, A. S., Anderson, D. G., Wharton, N. Short-term Comparison of Cervical

Fusion With Static and Dynamic Plating Using Computerized Motion Analysis. *Spine,* 2007; 32,13,371 - 5.

[48] Steinmetz, M. P., Benzel, E. C., Apfelbaum, I. R., Subsidence and dynamic cervical spine stabilization. In: Spine Surgery: techniques, complication avoidance and management. 2nd ed. Elsevier, 2005, Philadelphia.

[49] Khoo, L. T., Benae, J. L., Gravori, T. Anterior Plating for Cervical Traumatic Fractures: An Analysis of Graft Height and Segmental Lordosis Preservation: *The Internet Journal of Spine Surgery*, 2005; Volume 1, Number 2.

[50] Aryan, H. E., Sanchez-Mejia, R. O., Ben-Haim, S., Ames, C. P. Successful treatment of cervical myelopathy with minimal morbidity by circumferential decompression and fusion. *Eur. Spine J.,* 2007; 16(9):1401 - 9.

[51] Epstein, N. E. Fixed vs. dynamic plate complications following multilevel anterior cervical corpectomy and fusion with posterior stabilization. *Spinal Cord,* 2003; 41(7):379 - 84.

[52] Rapoff, A. J., O'Brein, T. J., Ghanayem, A. J. et al. Anterior cervical graft and plate load sharing. *J. Spinal Disord.,* 1999; 12:45 - 9.

[53] Saphier, P. S., Arginteanu, M. S., Moore, F. M., Steinberger, A. A., Camins, M. B. Stress-shielding compared with loadsharing anterior cervical plate fixation: a clinical and radiographic prospective analysis of 50 patients. *Journal of Neurosurgery*, (Spine) 2007; 6:391 - 397.

[54] Truumees, E., Demetropoulos, C. K., Yang, K. H., Herkowitz, H. N. Effects of a Cervical Compression Plate on Graft Forces in an Anterior Cervical Discectomy Model. *Spine,* 2003; 28(11):1097 - 1102.

[55] Cheng, B. C., Burns, P., Pirris, S., Welch, W. C. Load sharing and stabilization effects of anterior cervical devices. *J. Spinal Disord. Tech.,* 2009 Dec.; 22(8):571 - 7. doi: 10.1097/ BSD.0b013e31818eee78.

[56] Dvorak, M. F., Pitzen, T., Zhu, Q., Gordon, J. D., Fisher, C. G., Oxland, T. R. Anterior cervical plate fixation: a biomechanical study

to evaluate the effects of plate design, endplate preparation, and bone mineral density. *Spine,* 2005 30(3):294 - 301.

[57] Welch, W. C. Load sharing and stabilization effects of anterior cervical devices. *J. Spinal Disord. Tech.,* 2009 *J. Neurosurg. Spine,* 2008 Nov.; 9(5):444 - 9.

[58] Kirkpatrick, J. S., Levy, J. A., Carillo, J., Moeini, S. R. Reconstruction after multilevel corpectomy in the cervical spine. A sagittal plane biomechanical study. *Spine,* 1999; 24(12):1186 - 90; discussion 1191.

[59] Reidy, D., Finkelstein, J., Nagpurkar, A., Mousavi, P., Whyne, C. Cervical spine loading characteristics in a cadaveric C5 corporectomy model using a static and dynamic plate. *J. Spinal Disord. Tech.,* 2004; 17(2):117 - 22.

[60] Brodke, D. S., Gollogly, S., Alexander Mohr, R., Nguyen, B. K., Dailey, A. T., Bachus, A. K. Dynamic cervical plates: biomechanical evaluation of load sharing and stiffness. *Spine,* 2001 Jun. 15; 26 (12): 1324 - 9.

[61] Brodke, D. S., Klimo, P. Jr., Bachus, K. N., Braun, J. T., Dailey, A. T., Anterior cervical fixation: analysis of load-sharing and stability with use of static and dynamic plates. *J. Bone Joint Surg. Am.,* 2006 Jul.; 88 (7): 1566 - 73. and dynamic plates. *J. Bone Joint Surg. Am.,* 2006 Jul.; 88 (7): 1566 - 73.

In: A Closer Look at Biomechanics
Editor: Daniela Furst

ISBN: 978-1-53615-866-3
© 2019 Nova Science Publishers, Inc.

Chapter 4

A BIOMECHANICAL STUDY ON CORRELATION BETWEEN LATERALITY AND WALKING ASYMMETRY

Kadek Heri Sanjaya[*]
Research Center for Electrical Power and Mechatronics, Indonesian Institute of Sciences, Bandung, Indonesia

ABSTRACT

The development of laterality in humans has been associated with the evolution of bipedalism. The most observable laterality is the handedness, where around 85 to 90% of the population is right-handed. Another laterality features measured in this study is the footedness. The correlation between the handedness and footedness is unclear, especially among the left-handers. There are contradictory results from the previous studies on the effects of laterality on walking, such as the existence of symmetry or asymmetry as well as the role of the dominant leg. The effects of walking speed on walking symmetry are also not clearly understood. This article discusses the effects of laterality on walking asymmetry during walking on a treadmill. Participants of the walking

---

[*] Corresponding Author's E-mail: kade001@lipi.go.id.

experiment were seventeen healthy young adult males (11 right-handers and right-footers and 6 left-handers and mixed-footers, measured by Waterloo Handedness and Footedness Questionnaire). In the recorded anthropometry data, the right-handers showed acromion height discrepancy whereas left-handers showed the trochanteric height discrepancy. Both groups of participants showed biceps and lower thigh circumference discrepancy. Participants walked on a treadmill at 1.5, 3, and 4 km/h. Bilateral muscle activation was measured from the tibialis anterior, soleus, and lumbar erector spinae. Foot pressure sensors were attached bilaterally on five points of foot sole: big toe, first, third, and fifth metatarsals, and calcaneus. Cross-correlation function (CCF) analysis was employed to analyze any paired root mean square (RMS) of each bilateral muscle activation and foot pressure signals, which yielded CCF coefficient and time lag during one gait cycle to measure the symmetry. In general, our experiment showed that the left-handers had a greater asymmetry represented by lower CCF coefficient and longer time lag. In a further stage, CCF coefficient and time lag correlation with anthropometric data discrepancy were analyzed. The lower thigh circumference discrepancy, which ironically more correlated with handedness than footedness, was found to have the most consistent effects on the asymmetrical characteristics during walking as its effects were observed at all velocities measured. The greater asymmetry in left-handers probably related to their disadvantages in population study such as shorter life expectancy, higher safety risks, as well as musculoskeletal and sensory disorders. However, to what degree the disadvantages are embodied in physiological processes remain a big question necessary to be investigated in future studies with more appropriate experimental methods.

Keywords: electromyography, foot pressure, cross-correlation function, anthropometry discrepancy

INTRODUCTION

Symmetry is a phenomenon which represents the comparable quantity or quality of parts facing each other or around an axis so that the parts are composed into due proportion. Symmetry has been associated with equality, regularity, balance and good coordination. In the study of symmetry, there is a paradoxical phenomenon called as symmetry breaking where the symmetric system begins to act less symmetrically when the symmetry of the subsequent condition of the system is a subgroup of the entire symmetrical system (Stewart & Golubitsky, 2011). As shown in Figure 1, during walking, both legs move half a period out of phase which indicates symmetry breaking. In quadrupeds two types of symmetry are observable, namely spatial symmetry which shows the interchanging of fore and hind legs and spatio-temporal symmetry which shows the interchanging of left and right legs with a half-period phase shift, whereas in bipeds, only the latter is observable (Golubitsky, Stewart, Buono, & Collins, 1999). Figure 2 shows that in measuring gait, two important aspects are commonly used, namely gait cycle which represents the duration between foot-strike of the same leg and stance phase which represents the duration when the foot touches the ground (Stewart & Golubitsky, 2011).

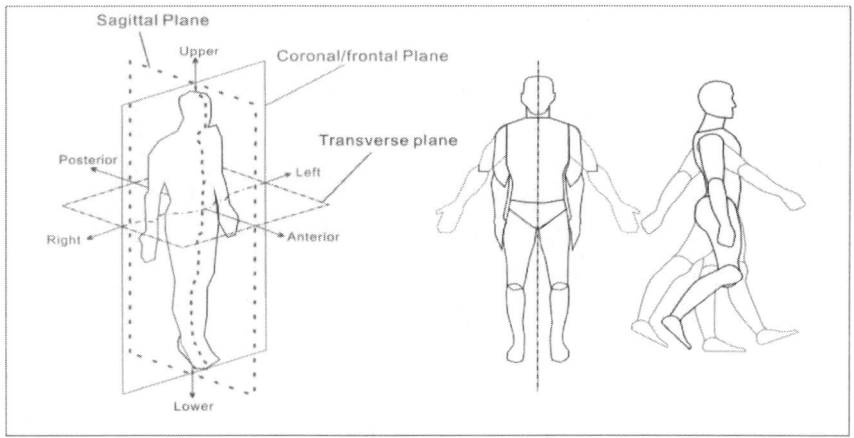

Figure 1. Symmetry axis of the human body that divides the body between the left and right side equally. During walking, bipedal humans walk with spatio-temporal symmetry.

One of the most common and observable asymmetrical things in human is laterality, which describes the asymmetrical preferential use of limbs and sensory (Schneiders et al., 2010). Handedness is the most familiar representative of laterality, followed by footedness. In evolutionary biology studies, the development of bipedalism has been closely linked with the development of laterality. Several previous studies on primates show the evidence of such an assumption. An observation on chimpanzees reported that a bipedal posture would stimulate hand preferences (Braccini, Lambeth, Schapiro, & Fitch, 2010), whereas a study on golden snub-nosed monkeys reported that foot preference is more observable in bipedal stance (Zhao, Li, & Watanabe, 2008). While right-handers domination is undeniable among humans, the exact percentage is varied. A study reported that 85% of the human population is right-handed (Uomini, 2009), while another study suggested that around 90% of humans are right-handed (Carey et al., 2001). Based on population studies, right foot tends to be the preferred foot of right-handers; however, the situation remains unclear for left-handers (Chapman, Chapman, & Allen, 1987). Therefore, the percentage of right-footers should be lower than right-handers, with the suggested percentage of around 80% of the population (Carey et al., 2001).

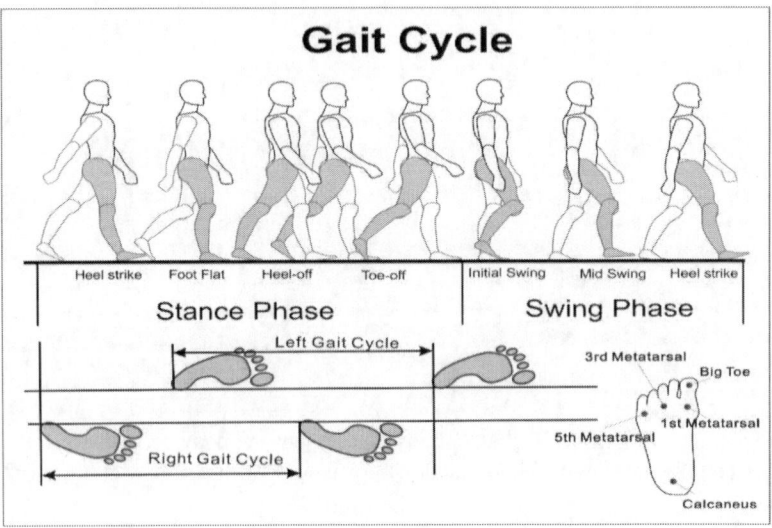

Figure 2. A gait cycle is marked by the period between two heel strikes. From the five pressure sensors attached on one foot, the pressure sensor on calcaneus is used to mark the gait cycle.

The left-footed subjects regulate their unipedal stance differently from the right-footers as footedness affects postural control (Golomer & Mbongo, 2004). However, lower limb choice for regulating posture in bipedal condition may be different from that in the unipedal, where the preferred leg is used for the more difficult task (Hart & Gabbard, 1997). Another study reported that during a static single-leg posture, footedness is not obvious, however higher frequency of movement strengthens lateral dominance for postural stability (Kiyota & Fujiwara, 2014). A study on footedness during walking using the footprint method found that footedness does not affect gait parameters (Zverev, 2009). Furthermore, gait initiation of normal subjects was reported to be highly symmetrical compared to the hemiparetic stroke subjects (Hesse et al., 1997). In contrast, a later study on normal subjects revealed that there is asymmetrical frontal body motion influenced by footedness in gait initiation (Dessery, Barbier, Gillet, & Corbeil, 2011). The contrasting results between the two studies were due to the different objectives and the subsequent different methods employed. The former focused on comparing the asymmetry between healthy subjects and hemiparetic patients gait parameters, as such, neglected laterality aspects. The researchers only used a non-parametric Wilcoxon test to measure the symmetry by comparing the bilateral gait parameters data collected by means of a force plate and video camera. The latter was intended to quantify asymmetry during gait initiation related to laterality by employing a more complicated measurement method which included electromyography (EMG), force plate, and Vicon motion analysis system followed by analyses using muscle coactivation index, the center of mass displacement, trunk inclination, and other gait parameters. A literature review concluded that asymmetrical lower limbs behavior during gait is a reflection of functional differences in propulsion and control (Sadeghi, Allard, Prince, & Labelle, 2000). A study using biomechanical method measured normal human gait using a symmetry index found that gait asymmetries were larger than expected (Herzog, Nigg, Read, & Olsson, 1989).

Stewart and Golubitsky (2011) hypothesized that the faster the movement, the more a system is stressed, the less symmetry it will become (Stewart & Golubitsky, 2011). The hypothesis is supported by a study which found greater propulsion on the dominant limb compared to non-dominant limb during fast walking (Seeley, Umberger, & Shapiro, 2008).

On the other hand, better limbs coordination to recover from perturbation was observed in faster walking but not associated with the gait stability (Krasovsky, Lamontagne, Feldman, & Levin, 2014). Furthermore, a study compared gait symmetry between primary school-aged children and young adults did not observe effects of speed on symmetry (Lythgo, Wilson, & Galea, 2011). In general, both symmetry and asymmetry are observed during gait; however, they cannot be generalized in association with the lateral dominance (Gundersen et al., 1989). The presumption of human body symmetry and the methods used to measure it determine the results of such studies. The contradictory is very well illustrated in anthropometry data recording. Most referred anthropometry data sources usually overlooked bilateral discrepancy, despite the fact that the limbs are largely asymmetrical in size, even when measured by a very simple method, as reported in a study conducted more than a century ago (Jones, 1870).

We assumed that these mixed results from previous studies are actually consistent with symmetry breaking concept (Stewart & Golubitsky, 2011), wherein dynamic systems, when symmetry becomes less stable and turns into asymmetry, there will be a reaction to recover the symmetry; thus both conditions are observed. Furthermore, symmetry breaking includes subgroups of the symmetrical system; therefore an analysis into smaller units such as measurement of various points of foot sole instead of global ground reaction force will be able to describe more detailed symmetry or asymmetry during gait.

To sum up, there were various studies on walking symmetry associated with the effect of laterality with contradictory results. Some studies suggested that walking among normal people are largely symmetrical, while some others suggested asymmetrical. Our experiment of walking while pushing showed that even walking among the healthy subjects are asymmetrical which is associated with laterality (Sanjaya, Lee, Shimomura, & Katsuura, 2014). Such condition was also observed in various walking condition (Sanjaya, Suherman, Lee, Shimomura, & Katsuura, 2017). In this study, we quantify the effects of laterality on walking asymmetry measured from the muscles activation and foot pressure. We also measured the laterality score, the grip strength between the left and right hands, as well as the anthropometry data bilateral discrepancy in order to analyze whether these factors represents laterality features that may influence walking asymmetry.

METHODS

Participants

In general, the methods used in this study has been described in our previous publication on manual pushing while walking on a treadmill (Sanjaya et al., 2014). This study required the identification of subjects' laterality before their inclusion in the experiment. Therefore we performed stratified random sampling when searching for participants. We looked for participants from the students population at Chiba University. The targeted participants should be healthy without any obvious signs of gait deficiency. The right-handers were selected randomly from a large group of right-handed students while the left-handers were selected from a specifically targeted group of students who use chopstick with left-hand. For this study, we had seventeen healthy young adult males participated as subjects (age 28 ± 5 years; height 169.9 ± 6.9 cm; weight 64.6 ± 7.3 kg), 11 of them were right-handers (50.25 ± 12.42) and right-footers (9.25 ± 4.13); and 6 of them were left-handers (-15.17 ± 13.7) and mixed-footers (0.0 ± 7.5) as assessed with Waterloo Handedness Questionnaire (WHQ) and Waterloo Footedness Questionnaire (WFQ) described in a reference (Elias, Bryden, & Bulman-Fleming, 1998). The results confirmed the suggestion that footedness among the left-handers are unclear as reported in the previous study (Chapman et al., 1987). Other than the laterality questionnaires, we also recorded bilateral anthropometry data of each subject from various body segments in order to observe their discrepancy as well as the handgrip strength. There were around 27 body dimensions of the participants measured in this study, as illustrated in Figure 3. To comply with the ethical approval required on the study involving human participants according to the Declaration of Helsinki, we had all subjects read and signed an informed consent form that had been approved by the ethical committee of Chiba University.

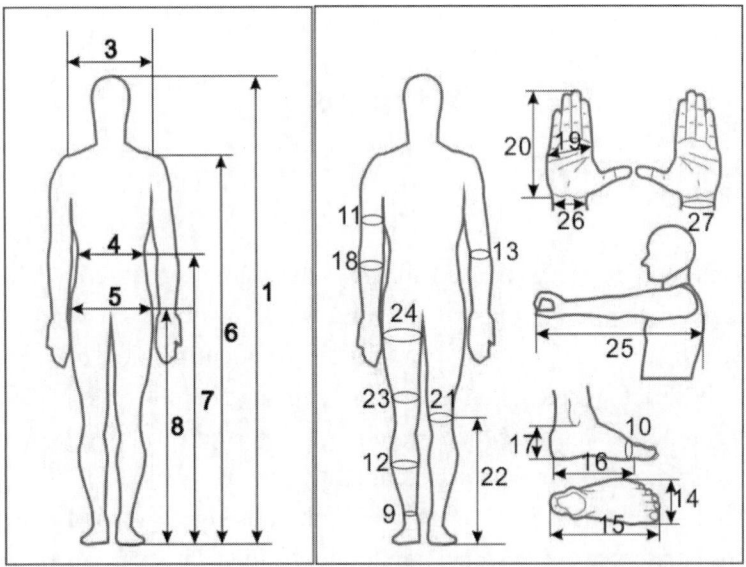

Figure 3. Body dimensions measured in this study to explore bilateral anthropometry data discrepancy among participants. Significant discrepancy observed on left-handers included trochanteric height (8), upper arm circumference (11), and lower thigh circumference (23). Significant discrepancy observed on right-handers included acromion height (6), upper arm circumference (11), and lower thigh circumference (23).

Instruments

Walking involves whole body movement. While the lower limbs movement has been the main focus of most human locomotion study, the hip mechanism has also been reported to prevent greater energy loss during walking (Kuo, 2001). As such, in this study, we recorded the muscle activation of lower limbs and lower back muscles. We attached surface electromyogram (EMG) electrodes bilaterally on the lower limb muscles, namely tibialis anterior and soleus, and lower back muscle of lumbar erector spinae. The muscles were identified using a palpation method as suggested by a reference (Muscolino, 2008). The tibialis anterior muscle plays a significant role in dorsiflexion of the ankle joint and assists in the inversion of the foot (Kendall, Florence Peterson, McCreary, Elizabeth Kendall Kendall, 1983). The soleus is one of the plantar flexor muscles that is activated during the toe-off phase in a gait cycle (Kendall, Florence Peterson, McCreary, Elizabeth Kendall Kendall, 1983). The erector spinae muscle is important in controlling trunk movement, such as lateral flexion, ipsilateral rotation of the trunk, and anterior trunk inclination by tilting and elevating the pelvis at the lumbosacral joint (Muscolino, 2008).

EMG electrodes were connected to a Biopac MP 150 data acquisition system (Biopac Systems, USA), which was further connected to a personal computer. Gait cycle was measured by the pressure sensors attached bilaterally on foot sole. Ten FSR-400 pressure sensors (Interlink Electronics, USA) were attached bilaterally on the big toe, first metatarsal, third metatarsal, fifth metatarsal, and calcaneus. The position of the pressure sensors on the foot sole is shown in Figure 2. The attachment was adapted from a method employed by Kiriyama et al. (2005) in order to measure foot pressure and contact duration (Kiriyama, Warabi, Kato, Yoshida, & Kokayashi, 2005). We had all of the subjects put on the same footwear. The onset and completion of each trial period were marked by a visual display timer and light sensor (Kodenshi Corp., Japan).

Experiment Procedure

Based on their effect on total mechanical energy (kinetic energy and potential energy), the forces in locomotion are divided into internal and external forces (P. A. Willems, Cavagna, & Heglund, 1995). Internal forces are the type of forces that cannot alter the total mechanical energy of an object, which includes among others gravity forces and spring force. The total mechanical energy of an object is constant when only internal forces that are doing work upon the object. There is only a change from potential energy into kinetic energy or vice versa; however, the total sum of both kinetic energy and potential energy do not change. External forces, on the other hand, are the type of forces that alter the total mechanical energy of an object, whose examples include normal force, friction force, and tension force. When external forces are performing work upon an object, the total mechanical energy of the object is going to change. The change can be positive or negative depends on whether there is energy gain or energy loss in the form of kinetic energy, potential energy, or both.

In this study, the subjects performed walking trials on a motorized treadmill (SportsArt Fitness, Taiwan). The setup speeds were 1.5 km/h, 3 km/h, and 4 km/h. Prior to the main trials, we allowed the subjects to train sufficiently with the walking condition on a moving platform such as a motorized treadmill. The walking speed of 3 km/h has been suggested to be the transitional speed with regard to internal work and external work comparison (Cavagna & Kaneko, 1977). At the walking speed slower than 3 km/h, the external work is greater than internal work, whereas at the walking speed faster than 3 km/h, the internal work is greater than the external work (Cavagna & Kaneko, 1977). The finding has also been confirmed by a later study (P. A. Willems et al., 1995). The duration of each walking trial was three minutes. For every walking speed, there were three trials performed. The subjects were given a three minutes rest between trials. While performing the trials, we instructed the subjects to look at the monitor which display timer of the trial session. Gazing on a monitor has been considered beneficial for this type of experiment because it prevents the occurrence of asymmetrical vision that may affect balance control as suggested in a previous study (Nagano, Yoshioka, Hay, Himeno, & Fukashiro, 2006).

Data Analysis

Data analysis was based on one gait cycle by using foot pressure sensor as a reference, where a gait cycle is a duration marked by the onset of a calcaneus-ground contact to the next contact as described in Figure 2. We divided the data into the left and right side gait cycle, where the data from left side muscles and pressure sensors were analyzed according to left foot gait cycle, and the data from the right side were analyzed according to right foot gait cycle. EMG and pressure sensor data were collected at a 1000-Hz sampling rate. Raw EMG signals were band-pass filtered between 15-250Hz, and root mean square (RMS) was derived. We selected RMS value because it represents the signal power and has a clearer physical meaning than integrated EMG (iEMG). In one gait trial of three minutes walking, we selected a series of 10 seconds of data in the middle of the session for further analysis. Furthermore, from the 10 seconds data, we selected 3 gait cycles for analysis. All data were normalized into 100 data points in one gait cycle.

The symmetry or asymmetry between the left and the right side were analyzed by comparing RMS of the recorded EMG signals and foot pressure data between left and right side using cross-correlation function (CCF) which yields time lag (τ) and cross-correlation function coefficient (CCF coefficient). The coefficients vary between -1 and +1, where a positive correlation value indicates the signals are in phase, and a negative value indicates inverse relationship (Nelson-Wong, Howarth, Winter, & Callaghan, 2009). Data measured from the left and right sides were arranged so that if the peak of right side were earlier than the left side, the time lag would be positive, while if it were later, the time lag would be negative.

In the statistical analyses, subjects were classified into one group for correlation analyses between laterality measures and biomechanical performance data, and two groups of the right-handers and left-handers for analyzing the differences between groups of handedness due to laterality. All data of each subject were averaged. We analyzed the characteristics of each group, and then for the change of CCF coefficient and time lag data, we also performed a comparison between both groups. Anthropometry discrepancy and hand grip strength data were calculated by subtracting the right value with the left value so that a positive result represents right side domination, a negative result represents the left side domination, and zero represents ambidexterity. These data then used further for correlation analysis.

The Shapiro-Wilk test was used to determine whether a parametric or non-parametric method should be used. Anthropometry discrepancy data correlation with WFQ and grip strength difference were analyzed using the parametric method of Pearson's product moment correlation coefficient while their correlation with WHQ was measured with the non-parametric method of Spearman's rank order correlation. Both the CCF coefficient and time lag correlation with anthropometry discrepancy were then analyzed by employing either Pearson's product moment correlation or Spearman's rank order correlation depends on the normal distribution of the data. We employed Student`s paired T-test for the parametric method and Wilcoxon signed-rank test for the non-parametric method in order to compare data between the left and right side. Data comparison between different walking speeds in the same group was performed using one-way ANOVA with repeated measures followed by Bonferroni post-hoc test for the parametric method and Friedman Test with the Wilcoxon signed-rank post-hoc test for the non-parametric method. Because the two groups of subjects have an unequal number of samples, in order to compare both groups, independent T-test for the parametric method and Mann-Whitney U test for non-parametric method were used. Statistical significance was set at $p < 0.05$. All statistical analyses were performed with Microsoft Excel 2010J and IBM SPSS 17J.

RESULTS

Anthropometry Discrepancy and Grip Strength

The data on subjects' anthropometric discrepancy has been published in our previous publication as we involved the same subjects for both experiments (Sanjaya et al., 2014). As described in Figure 3, we measured the dimension of 27 body dimensions. Paired T-test found that both right-handers and left-handers showed significant asymmetry in the upper arm circumference and lower thigh circumference ($p < 0.05$). The left-handers had greater left upper arm circumference (25.8 ± 1.8 cm) compared with their right upper arm (25.5 ± 1.9 cm). On the contrary, right-handers had smaller left upper arm circumference (26.3 ± 1.4 cm) compared with their right upper arm (26.6 ± 1.2 cm). The left lower thigh circumference of left-handers (39.4 ± 3.1 cm) was greater than the right side (39.0 ± 3.4 cm). The right-handers, on the other hand, showed smaller left lower thigh circumference (40.8 ± 2.7 cm) than their right side (41.5 ± 3.2 cm). The right-handers showed significant acromion height asymmetry ($p < 0.05$), where their left acromion (138.5 ± 7.5 cm) is taller than their right acromion (137.7 ± 7.1 cm). The left-handers did not show acromion height discrepancy; however, they showed significant asymmetrical trochanteric height ($p < 0.05$), where their left trochanter (86.8 ± 4.9 cm) is taller than their right trochanter (86.1 ± 4.8 cm). The other body segments dimension were found to be relatively symmetrical since no significant difference between the left and right side body segments were observed ($p > 0.05$).

The hand grip strength data were obtained by instructing the subjects to grip the hand dynamometer as strong as they could. The subjects performed the test several times, and two of the greatest data were averaged. The grip strength of left-handers did not show any significant differences between the left hand and right hand (left: 38.8 ± 6.0 kg, and right 38.7 ± 4.1 kg, $p > 0.05$). Different from the left-handers, the right-handers showed significantly weaker left-hand grip than the right hand (left: 37.6 ± 5.0 kg, and right 40.4 ± 4.0 kg, $p < 0.05$).

Gait Cycle Duration

Gait cycle duration was found to be shorter with faster walking in both left and right gait cycle of both groups of subjects ($p < 0.05$). The gait cycle duration of left-handed subjects were as follows: 1.49 ± 0.21 s (left) and 1.49 ± 0.20 s (right) at walking speed of 1.5 km/h, 1.19 ± 0.12 s (left) and 1.19 ± 0.12 s (right) at walking speed of 3 km/h, and 1.05 ± 0.07 s (left) and 1.05 ± 0.06 s (right) at walking speed of 4 km/h. The gait cycle duration of right-handers were as follows: 1.60 ± 0.38 s (left) and 1.60 ± 0.39 s (right) at walking speed of 1.5 km/h, 1.12 ± 0.11 s (left) and 1.12 ±

0.12 s (right) at walking speed of 3 km/h, and 1.00 ± 0.12 s (left) and 1.00 ± 0.12 s (right) at walking speed of 4 km/h. The asymmetry in the gait cycle duration in both groups was not observed since there was no difference between the left and right gait cycle duration ($p > 0.05$).

Foot-Ground Contact Duration

The data on the stance phase duration is shown in Table 1. The stance phase duration was observed to be relatively symmetrical both in the left- and right-handed subjects. In both groups of subjects, the increase in walking speed from 1.5 km/h to 4 km/h shortened stance phase duration significantly around 6% of the gait cycle in both feet ($p < 0.05$). The left-handers showed significantly shorter stance phase duration on left foot around 4% of the gait cycle when walking speed increased from 1.5 km/h to 3 km/h ($p = 0.02$). The right-handers showed significantly shorter stance phase duration on the left foot around 3% and 2% of gait cycle when walking speed increased from 1.5 km/h to 3 km/h ($p = 0.01$) and from 3 km/h to 4 km/h ($p = 0.005$) respectively, and on the right foot around 3% of gait cycle when walking speed increased from 1.5 km/h to 3 km/h ($p = 0.01$).

Table 1. Stance phase duration of both left- and right-handers, normalized into 100% of the gait cycle (mean ± SD)

Stance Phase Duration	Left-handers			Right-handers		
Speed	Left foot gait cycle	Right foot gait cycle	p	Left foot gait cycle	Right foot gait cycle	p
1.5 km/h	77.0 ± 2.5	74.5 ± 3.0	$p > 0.05$	74.8 ± 1.9	73.6 ± 3.2	$p > 0.05$
3 km/h	73.0 ± 1.2	71.0 ± 2.9	$p > 0.05$	71.5 ± 2.9	70.5 ± 3.2	$p > 0.05$
4 km/h	71.0 ± 2.2	68.5 ± 3.3	$p > 0.05$	69.1 ± 2.9	67.3 ± 2.7	$p > 0.05$

While the right-handed subjects did not show any asymmetrical contact duration in all the foot pressure sensors, the left-handers showed asymmetrical foot-ground contact duration. The asymmetry was observed on calcaneus at all speeds where the duration of the left calcaneus touched the ground significantly longer than the right calcaneus ($p < 0.05$). The calcaneus contact duration at walking speed of 1.5 km/h was 60.7 ± 8.9% for the left foot and 48.3 ± 11.1% for the right foot. At walking speed of 3 km/h, the contact duration of calcaneus was 48.8 ± 8.0% for the left foot and 36.3 ± 2.8% for the right foot. At the fastest walking speed of 4 km/h, the asymmetry was also observed where the contact duration of left calcaneus was 40.2 ± 5.4%, and that of the right calcaneus was 32.0 ± 3.2%. The contact duration of calcaneus at walking speed of 4 km/h was significantly shorter compared with the duration at 1.5 km/h ($p < 0.05$) in both the left and right foot. The asymmetry was also observed on the fifth metatarsal of the left-handers at 3 km/h walking speed ($p < 0.05$), where the contact duration of the left fifth metatarsal of 61.2 ± 3.1% was longer than its right counterpart at 54 ± 9.9%.

Foot Pressure Cross-Correlation Function Analysis

As shown in Figure 4, the left-handers showed a significant increase in CCF coefficient of big toe when walking speed increased from 1.5 km/h to 4 km/h (from 0.87 ± 0.06 to 0.93 ± 0.01; $p = 0.046$), indicating greater symmetry due to faster walking. The right-handers showed significantly higher CCF coefficient of big toe than left-handers at 1.5 km/h (left-handers: 0.93 ± 0.07, right-handers: 0.87 ± 0.06; $p = 0.026$) and 3 km/h (left-handers: 0.94 ± 0.05, right-handers: 0.87 ± 0.06; $p = 0.039$). The left-handers showed significantly longer time lag of big toe than the right-handers at 3 km/h walking speed (left-handers: 3.5 ± 3.2%, right-handers: 0.8 ± 1.3%, $p = 0.032$). Both lower CCF coefficient and longer time lag data suggesting that big toe pressure of the left-handers during walking is more asymmetrical than the right-handers.

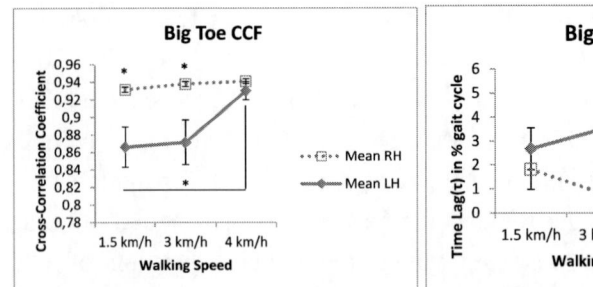
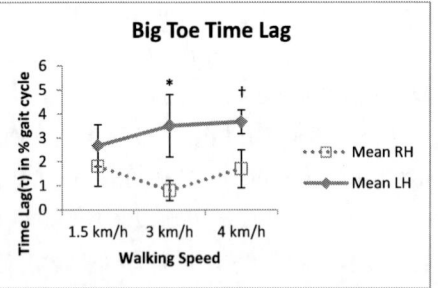

Figure 4. CCF coefficient and time lag of big toe (mean ± SE, $^\dagger p < 0.1$; $^* p < 0.05$). LH refers to left-handers, and RH refers to right-handers. Left-handers showed significantly lower CCF coefficient at 1.5 km/h and 3 km/h walking speed, and significantly longer time lag at 3 km/h walking speed.

As shown in Figure 5 (left), with regard to first metatarsal, the right-handers had significantly greater CCF coefficient of first metatarsal than their left-handed counterparts at 3 km/h (left-handers: 0.92 ± 0.02, right-handers: 0.95 ± 0.03, $p = 0.02$), suggesting the greater symmetry of right-handers compared with the left-handers with regard to first metatarsal pressure pattern on the ground. However, there was no asymmetrical timing of first metatarsal pressure as there were no significant differences observed in the time lag of the first metatarsal between groups of handedness and between speeds ($p > 0.05$).

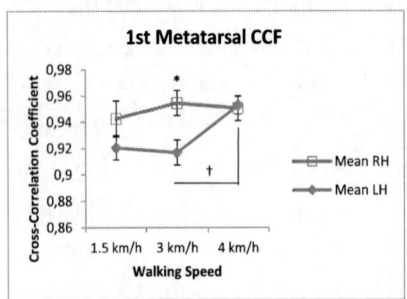
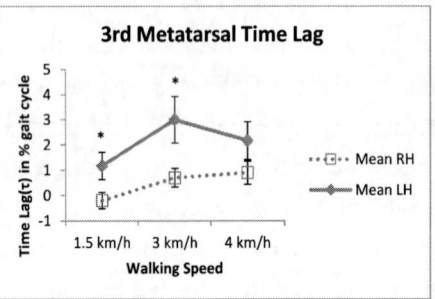

Figure 5. CCF coefficient of the first metatarsal and time lag of the third metatarsal (mean ± SE, $^\dagger p < 0.1$; $^* p < 0.05$). LH refers to left-handers, and RH refers to right-handers. Left-handers showed significantly lower CCF coefficient of the first metatarsal at 3 km/h and longer time lag of the third metatarsal at 1.5 km/h and 3 km/h.

There were no significant differences observed between groups of handedness and walking speeds in CCF coefficient of the third metatarsal ($p > 0.05$). Left-handers showed a significantly longer time lag than right-handers at the third metatarsal at 1.5 km/h (left-handers: 1.17 ± 1.33%, right-handers: -0.20 ± 1.03%; $p = 0.048$) and 3 km/h (left-handers: 3.00 ± 2.28%, right-handers: 0.70 ± 1.16%; $p = 0.022$). The time lag of the third metatarsal is described in Figure 5 (right). The data mean that despite the absence of asymmetrical third metatarsal pressure pattern, the timing of third metatarsal pressure is more asymmetrical compared with right-handers in that their right third metatarsal pressured the ground earlier than the left one in a gait cycle.

There were no significant differences observed in the CCF coefficient and time lag of the fifth metatarsal between groups of handedness and between walking speeds ($p > 0.05$). However, right-handers showed a tendency of more symmetrical walking as shown by the tendency of higher CCF coefficient at 3 km/h and 4 km/h as well as shorter time lags at both speeds compared to left-handers ($p < 0.1$).

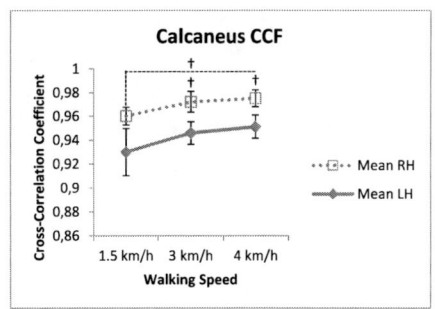
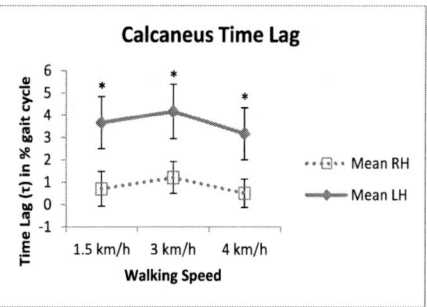

Figure 6. CCF coefficient and time lag of calcaneus (mean ± SE, $^\dagger p < 0.1$; $^* p < 0.05$). LH refers to left-handers, and RH refers to right-handers. Left-handers showed significantly longer time lag of calcaneus at 1.5 km/h, 3 km/h and 4 km/h.

Figure 6 shows that there were no significant differences observed between groups of handedness and walking speeds in the CCF coefficient of the calcaneus ($p > 0.05$). The left-handers showed significantly longer time lags than the right-handers at 1.5 km/h (left-handers: 3.67 ± 2.88%, right-handers: 0.70 ± 2.45%, $p = 0.045$), 3 km/h (left-handers: 4.17 ±

2.99%, right-handers: 1.20 ± 2.25%, $p = 0.040$) and 4 km/h (left-handers: 3.17 ± 2.86%, right-handers: 0.50 ± 2.01%, $p = 0.045$). Just like what was observed on the third metatarsal, the data suggested that the timing of calcaneus pressure is more asymmetrical compared with right-handers in that their right calcaneus pressured the ground earlier than the left one in a gait cycle. These data are probably related to the asymmetrical calcaneus contact duration of left-handed subjects as reported in the previous section.

Muscle Activation Cross-Correlation Function Analysis

Tibialis anterior muscle activation did not show any significant effects of laterality and walking speed in both CCF coefficient and time lag ($p > 0.05$). Right-handers only showed a tendency of greater CCF coefficient when walking speed increased from 1.5 km/h to 4 km/h ($p = 0.074$).

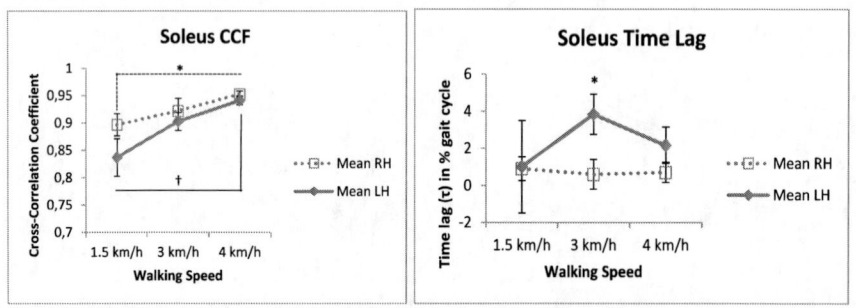

Figure 7. CCF coefficient and time lag of soleus muscle (mean ± SE, $^{†}p < 0.1$; $^{*}p < 0.05$). LH refers to left-handers, and RH refers to right-handers. Right-handers showed increasing CCF coefficient when walking speed was increased from 1.5 km/h to 4 km/h. Left-handers showed a longer time lag at 3 km/h.

Figure 7 shows that faster walking speed from 1.5 km/h to 4 km/h resulted in significant increase of the soleus muscle activation CCF coefficient in right-handers (from 0.90 ± 0.06 to 0.95 ± 0.02; $p = 0.011$), but not in left-handers ($p > 0.05$). The data suggested that faster walking speed improves the symmetry on the right-handers but not the left-handers. The left-handers showed longer time lag of soleus muscle activation than right-handers at 3 km/h walking speed (left-handers: 3.83 ± 2.64%, right-handers: 0.60 ± 2.41%, $p = 0.025$). The data indicated that at walking speed of 3 km/h, soleus muscle of the right foot in the left-handed subjects activated earlier than its counterpart in the left foot, and the bilateral difference in activation timing was more staggering among the left-handed subjects than the right-handed subjects.

As shown in Figure 8, the right-handers showed a significant increase of erector spinae muscle activation CCF coefficient from 1.5 km/h to 4 km/h walking speed (from 0.78 ± 0.11% to 0.91 ± 0.02%; $p = 0.007$), indicating that faster walking speed improved symmetrical activation between left and right lumbar erector spinae. This increase also resulted in a significant difference between left-handers and right-handers in erector spinae muscle activation CCF coefficient at 4 km/h walking speed (left-handers: 0.78 ± 0.10%, right-handers: 0.91 ± 0.02%, $p = 0.005$), suggesting that the left-handers low back muscle such as lumbar erector spinae was activated more asymmetrically than the right-handers. Significant time lags differences were not observed in all conditions measured ($p > 0.05$).

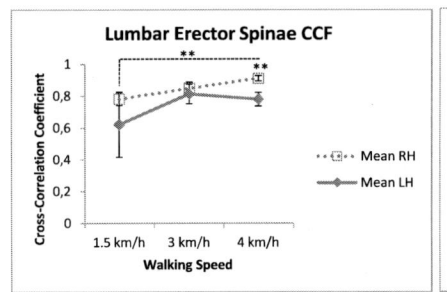 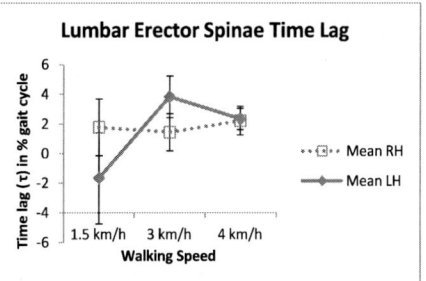

Figure 8. CCF coefficient and time lag of soleus muscle (mean ± SE, **$p < 0.01$). LH refers to left-handers, and RH refers to right-handers. Right-handers showed a significant increase of erector spinae muscle CCF coefficient with faster walking from 1.5 km/h to 4 km/h. Left-handers showed significantly lower erector spinae CCF coefficient at 4 km/h walking speed.

Correlation Analysis

Handedness has a significant positive correlation with both upper arm ($r_s = 0.559$, $p < 0.05$) and lower thigh circumference discrepancy ($r_s =$

0.605, $p < 0.05$) whereas leg length difference as represented by the trochanteric height discrepancy showed a significant positive correlation with footedness score ($r = 0.565$, $p < 0.05$). Grip strength did not show significant correlation with any asymmetrical anthropometry data measured ($p > 0.05$).

Acromion discrepancy and trochanteric height did not affect both CCF coefficient and time lag of all muscle activation and foot pressure measured ($p > 0.05$). Upper arm circumference discrepancy which is correlated with the degree of handedness has limited effects on walking asymmetry, at the lowest speed of 1.5 km/h, it has a negative correlation with the time lag of the first metatarsal ($r_s = -0.522$, $p = 0.038$) and a positive correlation with erector spinae CCF coefficient at walking speed of 4 km/h ($r_s = 0.594$, $p = 0.019$).

The most dominant anthropometry measure which affects walking asymmetry is lower thigh circumference. The effects of lower thigh circumference discrepancy on walking asymmetry were observed almost at all speeds measured, on both CCF coefficient and time lag. At 1.5 km/h, the lower thigh circumference asymmetry has a significant positive correlation with CCF coefficient of the big toe ($r = 0.541$, $p = 0.030$), and a significant negative correlation with time lag of the third metatarsal ($r_s = -0.528$, $p = 0.035$) and the calcaneus ($r = -0.602$, $p = 0.014$).

In general, at 3 km/h, lower thigh circumference also showed significant positive correlation with CCF coefficient of the first metatarsal ($r = 0.608$, $p < 0.05$) and calcaneus (0.496, $p < 0.05$). With regard to time lag, it has a significant negative correlation with the big toe ($r = -0.648$, $p < 0.01$), third metatarsal ($r = -0.614$, $p = 0.011$), calcaneus ($r = -0.588$, $p = 0.017$), and soleus muscle activation ($r = -0.720$, $p < 0.01$).

Just like what had been observed on walking at 1.5 km/h and 3 km/h, at 4 km/h, the lower thigh asymmetry also shows a positive correlation with CCF coefficient and negative correlation with the time lag. The lower thigh circumference discrepancy has significant negative correlation with the CCF coefficient of fifth metatarsal ($r = 0.617$, $p = 0.011$) and calcaneus ($r_s = 0.659$, $p < 0.01$). The lower thigh circumference discrepancy has significant negative correlation with time lag of the big toe ($r = -0.566$,

$p < 0.05$), first metatarsal ($r_s = -0.520$, $p < 0.05$), third metatarsal ($r_s = -0.620$, $p < 0.05$), and soleus muscle activation ($r = -0.589$, $p < 0.016$).

DISCUSSION

Laterality is probably still one of the greatest mysteries of human life. The origin of laterality has been studied extensively both through fossils or comparative studies with non-human primates; however, the results remain inconclusive. The fossils studies reported that the strongest evidence of handedness was observed among the Neanderthal population (Cashmore, Uomini, & Chapelain, 2008). The fossils unearthed from the Neanderthal population indicated left brain dominance which is signaling linguistic competence. The fossils also indicated asymmetrical arms movement as observed on the traces preserved on the teeth (Volpato et al., 2012). The studies based on non-human primates observation concluded that handedness data are determined by the context, such as left-handed dominance was reported among males chimpanzees in an observation in Tanzania, while the same observation found that females population were more right-handed (Corp & Byrne, 2004), and the dominant of right-handedness was also observed among captive orang-utans (Hopkins, Stoinski, Lukas, Ross, & Wesley, 2003). On the other hand, left-hand preference in order to reach objects such as foods was reported in lemurs, rhesus monkeys, and Japanese macaques (Hopkins et al., 2011). There is speculation that arboreal primates have a greater tendency toward left-hand preferences whereas terrestrial primates population have a greater tendency to be dominated by right-handedness (Hopkins et al., 2011). However, basically, research on non-human primates showed inconsistent patterns across different populations and environments (Uomini, 2009).

In the study of evolution, the emergence of laterality has been associated with bipedalism. The necessity to carry loads such as food, tools or infants was speculated to be a stimulus for bipedalism (Duarte, Hanna, Sanches, Liu, & Fragaszy, 2012), while bipedalism observed on non-human primates have been reported to strengthen laterality. Studies on chimpanzees (Braccini et al., 2010), Sichuan snub-nosed monkeys (Zhao et al., 2008), and Japanese macaques (Leca, Gunst, & Huffman, 2010) suggested that bipedal stance evoke greater hand or foot preferences. However, different from non-human primates, the cerebral cortex of modern humans is characterized by certain functional asymmetry. The asymmetry is an evolutionary consequence of asymmetric employment of the forelimbs coupled with linguistic mechanisms (Frost, 1980). The statement was supported by earlier research on the influence of tool-use on brain development (Beck, 1974). In general, despite the extensive research, the timing of laterality emergence in the evolution phases remains unclear, and various methods such as comparison from fossils or other great apes in existence have been producing inconsistent results (Cashmore et al., 2008).

Our knowledge on the origin of laterality and the reason for its existence is limited. However, the left-handers have been living in the human population since the beginning of any known civilization with their own well-documented problems. The left-handers have been subjected to prejudice, humiliation, and discrimination in many societies (Coren, 1992). There are many pieces of evidence of unconsciously negative perception and attitudes toward left-handers. The meaning of the word left in various languages is negatively perceived words such as weak, broken, awkward, clumsy, ugly, and wrong (Coren, 1992). Furthermore, in many religions and beliefs, left-handedness has been associated with the devil.

Population studies put forward several data on the disadvantages of left-handedness as described thoroughly in a reference book (Coren, 1992). The percentage of left-handers in the population with psychological problems was reported to be higher than in a healthy population. The born left-handers are associated with various childbirth problems and difficulties. With regard to accidents during performing activities such as sport, work, home, using tools, and driving, the left-handers have a much higher probability of getting accidents between 20% to more than 80%. Based on baseball players data and Southern California death certificate investigation, the left-handers were found to die sooner than their right-handed counterparts, especially more staggering among the male subjects where the shortened life span is twice than that of the female ones. After all the mentioned drawbacks of left-handedness, the relevant question with this study should be whether such disadvantageous condition exists in gait and whether it can be measured biomechanically.

Measurement using WFQ and hand grip strength showed that the left-handers were more ambidextrous than the right-handers as they had mixed footedness and insignificant differences between left and right-hand grip strength. However, the results of the walking experiment in this study suggest that the left-handers walk more asymmetrical than the right-handers. In general stance phase duration, asymmetry was not observed both in left- and right-handers, however, when we measured in further detail, we found that the asymmetries were observed on both on the big toe, metatarsal heads, and calcaneus especially in the left-handers. The advantages offered by the method used in this study is the ability to reveal these asymmetries on several points on the foot sole and muscle activation.

In a previous study, Lambrinudi described the main function of the big toe is to support the metatarsal heads to take the full body weight when the heel is raised (Lambrinudi, 1932). The range of motion of the big toe lies in dorsiflexion which occurs twice during the stance phase of walking, namely active and passive dorsiflexion (Bojsen-Møller & Lamoreux, 1979). The active dorsiflexion occurs right before the calcaneus contacts the ground and continues until after the metatarsal heads contact (Bojsen-Møller & Lamoreux, 1979). The passive dorsiflexion starts after the calcaneus takes off the ground prior to push-off, as the toes are forced dorsally by the weight of the body (Bojsen-Møller & Lamoreux, 1979). In this study, foot pressure sensors only measured the passive big toe dorsiflexion. Other than body weight bearing, the big toe dorsiflexion was also hypothesized to have an effect on venous flow in the metatarsals head of foot as it presses the cutaneous and subcutaneous venous plexus, which is well developed in the ball of the foot (Bojsen-Møller & Lamoreux, 1979).

The left-handers showed more asymmetrical big toe pressure than right-handers as indicated by lower CCF coefficient and longer time lag especially at 3 km/h. The time lags of big toe were positive in both left- and right-handers, therefore the pressure of the right big toe was earlier than the left big toe. As big toe dorsiflexion has an effect on venous flow, we may speculate that there should be asymmetry on the flow, and this assumption should be a subject for future research. In static standing, a study with a large number of subjects from preschool children reported slightly greater anterior foot pressure in the right foot, indicating asymmetry exists in a static posture; however, the study did not identify the laterality of the subjects (Matsuda & Demura, 2013). Furthermore, the previous study reported that they did not observe such asymmetry among adult subjects, and was unable to describe the reason of asymmetry among children as the study was not intended to investigate laterality (Matsuda & Demura, 2013). Our study, despite the different experiment method, has the advantage that we prepared to measure left and right difference from the beginning, and as such, probably capture such phenomenon of the asymmetry with faster right big toe foot pressure than the left one both in left- and right-handers. However, the time lags of the left-handers were around three times longer than those of the right-handers, suggesting that despite they are mixed-footers, how they manage the timing of the pressure of their big toes as weight-bearer during push off are much more asymmetrical. Laterality measures in the form of WHQ and WFQ did not show any correlation with this asymmetry. On the other hand, the lower thigh circumference discrepancy showed a positive correlation with CCF coefficients of the big toe, furthermore supporting the assumption that the left-handers had more asymmetrical lower limbs weight-bearing strategy in the form of big toe pressure pattern. Such strategy was not observed in the timing of the pressure, since there is no significant correlation between the lower thigh asymmetrical anthropometry and the time lag data, suggesting body mass of the human body parts probably has an effect on magnitude and pattern of the pressure but not the timing.

A previous study reported that the big toe, the first and second metatarsals took around 64% of the total anterior load during the push-off phase in a gait cycle (Hayafune, Hayafune, & Jacob, 1999). They also found that body weight bearing of big toe has negative correlations with body weight bearing of the third metatarsal, but not first and fifth metatarsals (Hayafune et al., 1999). When foot touches the ground, the foot pronates and then supinates (Hutton & Dhanendran, 1979), which is indicated by the first metatarsal touches the ground earlier than both third and fifth metatarsals. The supination distributes body weight pressure from the first metatarsal laterally, as evidenced by a negative correlation of medio-lateral load in the previous study (Hayafune et al., 1999).

While no significant differences were observed on the fifth metatarsal, right-handers showed significantly greater CCF coefficient on the first metatarsal at 3 km/h than left-handers, indicating more symmetrical pressure pattern, followed by a shorter time lag on the third metatarsal at 1.5 and 3 km/h, indicating a greater coincidence in the pressure timing of the third metatarsal of both feet. The time lag of left-handers at both 1.5 and 3 km/h were both positive, indicating the peak pressure of the right third metatarsal is earlier than the left third metatarsal. In the correlation analysis to find the effect of asymmetrical anthropometry data, only the lower thigh circumference showed an effect on the time lag of both big toe and the third metatarsal at 3 and 4 km/h walking speed, in the form of negative correlation. This was different from its positive correlation with CCF of the big toe and negative correlation with the time lag of the third metatarsal at 1.5 km/h, therefore indicating the correlation between the big toe and the metatarsal heads changes due to walking speed, in disagreement with the previous study (Hayafune et al., 1999). The right-handers had larger lower thigh circumference of their right leg, whereas the left-handers' data were on the contrary in that their left lower thigh circumference was larger. However, the effects on asymmetrical anterior foot pressure were much more staggering among the left-handers. The results suggest that the strategy of the left-handers when the big toe and metatarsals touch the ground from the end of heel strike to push-off is more asymmetrical. This asymmetry occurs across the big toe dorsiflexion and the pronation and supination of the metatarsal heads. We speculate there should be greater asymmetrical pressure on the cutaneous and subcutaneous venous plexus at the metatarsal heads of the left-handers.

Calcaneus which is the largest tarsal bone covers the posterior area of the foot. This area touches the ground earlier than the anterior foot, especially during heel strike at the onset of the stance phase. The role makes it naturally able to withstand high tensile, bending and compressive forces, however high instantaneous loads that occur during heel-strike may lead to fracture (Hall & Shereff, 1993). Surprisingly, the magnitude of loading on calcaneus do not peak at high impact at the onset of heel strike, but at the later stage of stance phase, especially around 60% of stance phase or right after the mid-stance (Giddings, Beaupre, Whalen, & Carter, 2000).

In this study, we found that the left-handers had longer time lag of calcaneus significantly at all speeds measured. Both left- and right-handers showed positive time lags indicating the right calcaneus achieved the peak pressure earlier than the left calcaneus. When we checked the stance phase duration of the left and right calcaneus among the left-handers, we found that the left calcaneus stance duration was significantly longer than the right calcaneus. Such left and right difference was not observed on the right-handers. This condition may indicate the existence of left and right legs role difference with regard to the propulsion-control hypothesis, where the right foot is used for propulsion and the left foot is used for control. The left-handers with greater problem of asymmetry may require longer left-foot support for maintaining balance. Greater asymmetrical loading both in magnitude and the instantaneous timing on calcaneus may lead to greater risk of fracture. Previous studies based on population data reported that left-handedness had been associated with a greater risk of fracture at pelvis and limbs bones (Luetters, Kelsey, Keegan, Quesenberry, & Sidney, 2003). In the soft tissues that envelope calcaneus, at the lateral side to the calcaneus have been associated with the areas of skin necrosis that may be related to the arterial anatomy (Hall & Shereff, 1993). As such, we speculate that the asymmetry observed on calcaneus may also have an effect on blood circulation.

Tibialis anterior muscle has an important role in ankle dorsiflexion (Kendall, Florence Peterson, McCreary, Elizabeth Kendall Kendall, 1983). Tibialis anterior, as part of leg flexor muscles, is characterized by high responsiveness to visual stimuli and stronger cortico-spinal projections to lower leg motoneuron (Van Hedel, Biedermann, Erni, & Dietz, 2002). In this experiment, the tibialis anterior muscle did not show any significant differences in all condition measured. We controlled the visual stimuli and the gaze of subjects so that those aspects would not affect balance perception. The only visual stimuli given were to mark the start and finish of each trial session.

Soleus muscle has a role in ankle plantar flexion (Kendall, Florence Peterson, McCreary, Elizabeth Kendall Kendall, 1983). Different from the tibialis anterior, soleus, which is an extensor muscle, characterized by greater responsiveness to somatosensory input (Van Hedel et al., 2002). Furthermore, together with gastrocnemius muscle, soleus muscle is responsible for about 93% of plantar flexor torque during normal gait (Giddings et al., 2000). In our experiment, soleus muscle showed a significant difference between the left- and right-handers in time lag at 3 km/h, where the left-handers had significantly longer positive time lag. Soleus muscle of the right-handers also showed a significant increase of CCF coefficient between 1.5 and 4 km/h, suggesting that faster walking improve symmetry. In this study, subjects performed walking on a motorized treadmill that unavoidably provided rhythmic somatosensory cueing that may affect soleus muscle activation patterns. Such an effect has been reported in a study on Parkinson patients to improve the gait stability (A. Willems et al., 2006).

The improved gait stability as indicated by greater symmetry in this study was only observed among the right-handers. Furthermore, observation on interlimb coordination during walking indicates that ankle dorsiflexor half centers of homologous limbs inhibit each other, whereas the plantar flexor half centers are not coupled each other (Van Hedel et al., 2002), therefore we may assume that the proprioceptive information continuously modulates soleus muscle activation but not tibialis anterior muscle which is controlled by central inputs. The plantar flexors such as soleus muscle have been reported to be important during support, forward progression and swing initiation, and also contributes to anterior-posterior and vertical ground reaction force during propulsion phase (Neptune & Sasaki, 2005). The left-handers were found to have more asymmetrical plantar flexor muscle activation which was not improved with regard to symmetry by the increase of walking speed and rhythmic somatosensory cueing. The left-handers have been reported to have greater risks to suffer sensory disorder (Coren, 1992), a situation which may be related to the unaffected asymmetrical soleus muscle activation. Lower thigh circumference discrepancy affected the time lag of soleus muscle activation in a negative correlation at 3 and 4 km/h walking speed suggesting that thigh mass affect soleus muscle activation timing during the shifting to greater internal force mechanism when the change from potential energy into kinetic energy or vice versa is getting more efficient.

The erector spinae muscle was found to have greater symmetry due to faster walking in right-handers as indicated by the increase of CCF coefficient from 1.5 km/h to 4 km/h. The right-handers also showed a greater symmetry than the left-handers at 4 km/h as proven by the significantly higher CCF coefficient. The increase of erector spinae CCF coefficient resembles that of soleus muscle in the right-handed subjects. In walking, the peak activation of erector spinae were found between contralateral heel strike and the onset of swing phase of ipsilateral leg, almost the same time with soleus muscle whose peak activation lies between the push-off and onset of the swing phase of ipsilateral leg which is in agreement with the previous studies (Ivanenko, Poppele, & Lacquaniti, 2004). Erector spinae muscle main function during walking is to maintain postural stability by restricting excessive trunk movement especially in the frontal plane (Thorstensson, Carlson, Zomlefer, & Nilsson, 1982). The more asymmetrical erector spinae muscle activation among the left-handers suggests that they have worse postural stability that may be accompanied by more uncontrolled trunk movement.

Despite very limited reference that directly points the association of the left-handedness and low back pain, musculoskeletal disorders such as the development of thoracic hyperkyphosis during growth have been reported to be more prevalent among the left-handers (Nissinen, Heliövaara, Seitsamo, & Poussa, 1995). Thoracic hyperkyphosis may alter the body center of gravity and therefore also alter the balance and body coordination during gait, especially with the tendency of greater asymmetry among the left-handers. Furthermore, a posturally induced stretch weakness on hip abductor muscles due to laterality and daily habits in asymmetrically carrying loads has been reported to affect symmetry (Neumann, Soderberg, & Cook, 1988). While we did not perform a bilateral muscle strength measurement in this study, the significant trochanteric height discrepancy that was only observed on the left-handed subjects probably confirm the above-mentioned condition. Furthermore, as hip mechanism plays a significant role to prevent greater energy loss during walking, we may speculate that greater asymmetry on hip and lower back area among the left-handers probably an indication of their greater energy loss.

Bipedal humans walk with erect trunk on two legs, that is naturally unstable especially medio-laterally; therefore the development of lower limb movement and trunk vertical stability emphasizes the dynamic coordination of body balance and forward motion (Courtine, Papaxanthis, & Schieppati, 2006). The human's foot is characterized with narrow width and longer length, and less stable compared to non-human primates such as chimpanzee which has a wider width and abducted big toe (Kiriyama et al., 2005). However, this foot shape with an arc structure allows speedy progressions at the expense of medio-lateral stability (Kiriyama et al., 2005). As a study reported that limb dominance did not affect the lower limbs kinematic and kinetic patterns during running (Brown, Zifchock, & Hillstrom, 2014), it seems that faster gait will increase the gait symmetry, especially for the right-handers. The greater symmetry in faster walking is in disagreement with the previous hypothesis that assumes faster walking will result in more asymmetry as the system suffers greater stress (Stewart & Golubitsky, 2011). However such improved symmetry phenomenon was not observed on the left-handers. By taking into account the principle of whole body coordination, an alteration on a certain part of the human body will eventually affect the other body parts. Special treatment on foot, especially between toes and metatarsals head, was also reported to have an effect on erector spinae muscle activation since foot pronation affected the internal rotation of the leg and ipsilateral pelvic tilt (Bird, Bendrups, & Payne, 2003). We speculate that the more asymmetrical foot pressure on the big toe and metatarsal heads, followed by asymmetrical soleus muscle activation and significant asymmetry on trochanteric height affect the more asymmetrical walking in the left-handed subjects. However, we should avoid oversimplification that those aspects themselves are the main cause of the asymmetrical walking as locomotion is a complex operation involving the coordination of the whole human body.

Animal locomotion to a limited degree is governed by an intraspinal network of neurons capable of generating rhythmic movements of limbs which are called as central pattern generators (CPG) (Golubitsky et al., 1999). While the CPG model has been considered to be symmetrical (Golubitsky et al., 1999), in the long evolutionary process, skilled hand movements as an adaptation for tool use probably evoked greater role of the direct cortico-motoneuronal system. The greater cortico-motoneuronal system interference furthermore affects the degree of laterality in humans and non-human primates (Dietz, 2002). Greater symmetry as a consequence of the faster walking speed as observed especially in soleus and erector spinae muscles of the right-handed subjects probably related to the higher frequency of rhythmical movement. The faster rhythmical movement indicates a greater responsibility for the CPG. However such effect was not observed in the left-handers. Human locomotion is also controlled by the central nervous system (CNS) by putting to use a series of activation patterns. These activation patterns are distributed to several different muscles which produced output during phases of motor task based on both feedforward and feedback signals from the dynamic condition of the whole limbs (Lacquaniti, Ivanenko, & Zago, 2012). We speculate that there is a difference in the degree of direct cortical-motoneuronal and intraspinal intervention during locomotion between the left- and right-handers.

The most common obstacle in the studies of handedness is the limited number of non-right-handed subjects; therefore generalization from the right-handed subjects has been a frequent method applied. The results of our present study show that such a generalization is irrelevant to the left-handed and mixed-footed subjects whose asymmetrical features are different from the right-handed subjects. Laterality is not constant from the birth and about 10% of each left- and right-handers were found switched into the opposite handedness (Coren, 1992). During walking, bilateral muscle activation patterns were observed (MacLellan et al., 2014) and asymmetrical interlimb learned ankle movement was reported (Morris, Newby, Wininger, & Craelius, 2009). The whole body coordination, just like other body mechanisms, is attached to the dynamic state of equilibrium principle. One embodiment of the equilibrium is symmetry. When there is a tendency of disturbance to the equilibrium, the human body will react and make an adjustment to preserve balance. This study found that the left-handers walk more asymmetrically than the right-handers; as such, they may be subjected to a harder effort to preserve balance. The condition may eventually have consequences that many researchers on population studies measured, such as higher risk of accident, musculoskeletal disorders, as well as common problems in physiology and psychophysiology. This study needs to be replicated in a study involving a large number of subjects. Furthermore, more sophisticated techniques to investigate the association between the physiological condition of the left-handers and the recorded consequences in population studies are necessary as the results of this study open the possibility for such studies. In order to draw a general conclusion on laterality during human evolution, the availability of various subjects is important, and future study should avoid generalizing any finding in right-lateralized subjects in other laterality groups.

CONCLUSION

This study was not intended to answer the mystery related to the origin of laterality, as it is a difficult question. The investigation of such a big research question requires enormous skill and effort. Considering our research team capacity and the availability of research methods at the present time, answering such a question will be out of reach. This study explored the disadvantages of left-handedness during walking from a biomechanical perspective. The left-handed subjects were found to have more asymmetrical foot pressure and muscle activation. Unlike the right-handers, faster walking speed did not improve the symmetry of the left-handers. The asymmetry probably has effects on less efficient energy management during walking as a greater effort required to sustain balance. As muscle activation and generating force during gait are closely related to the blood circulation, the asymmetry could have cardiovascular effects; as such, this topic should be subject for future study. The population studies reported the disadvantages of left-handedness such as shorter life expectancy, higher safety risks, as well as musculoskeletal, sensory, and sleep disorders. However, to what degree the disadvantages are embodied in physiological processes remain a big question to be investigated. More focused studies investigated the correlation between the laterality and human physiology with a larger number of participants are necessary to perform in the future in order to greatly improve our understanding of laterality and its consequences.

ACKNOWLEDGMENTS

The early experiment stage of this study was performed at Chiba University, Japan and was partially funded by the Monbukagakusho (MEXT) Government to Government scholarship program. Further stage of the experiment, as well as the data analysis and publication of this study, were performed at the Research Center for Electrical Power and Mechatronics, Indonesian Institute of Sciences (RCEPM-LIPI) in Bandung, Indonesia and supported by the Ministry of Research, Technology and Higher Education of the Republic of Indonesia

(Kemenristekdikti) through INSINAS research funding scheme with the contract number 15/INS-1/PPK/E4/2019.

REFERENCES

Beck, B. B. (1974). Baboons, chimpanzees, and tools. *Journal of Human Evolution*, *3*, 509–516. http://doi.org/10.1016/0047-2484(74)90011-6.

Bird, A., Bendrups, A., & Payne, C. (2003). The effect of foot wedging on electromyographic activity in the erector spinae and gluteus medius muscles during walking. *Gait and Posture*, *18*, 81–91.

Bojsen-Møller, F., & Lamoreux, L. (1979). Significance of free dorsiflexion of the toes in walking. *Acta Orthopaedica*, *50*(4), 471–479. http://doi.org/10.3109/17453677908989792.

Braccini, S., Lambeth, S., Schapiro, S., & Fitch, W. T. (2010). Bipedal tool use strengthens chimpanzee hand preferences. *Journal of Human Evolution*, *58*(3), 234–41. http://doi.org/10.1016/j.jhevol.2009.11.008.

Brown, A. M., Zifchock, R. A., & Hillstrom, H. J. (2014). The effects of limb dominance and fatigue on running biomechanics. *Gait and Posture*, *39*(3), 915–919. http://doi.org/10.1016/j.gaitpost.2013.12.007.

Carey, D. P., Smith, G., Smith, D. T., Shepherd, J. W., Skriver, J., Ord, L., & Rutland, a. (2001). Footedness in world soccer: an analysis of France '98. *Journal of Sports Sciences*, *19*(11), 855–864. http://doi.org/10.1080/026404101753113804.

Cashmore, L., Uomini, N., & Chapelain, A. (2008). The evolution of handedness in humans and great apes: A review and current issues. *Journal of Anthropological Sciences*, *86*, 7–35.

Cavagna, G. A., & Kaneko, M. (1977). Mechanical work and efficiency in level walking and running. *Journal of Physiology*, *268*, 467–481.

Chapman, J. P., Chapman, L. J., & Allen, J. J. (1987). The measurement of foot preference. *Neuropsychologia*, *25*(3), 579–584. http://doi.org/10.1016/0028-3932(87)90082-0.

Coren, S. (1992). *The Left-Hander Syndrome: The Causes and Consequences of Left-Handedness* (1st ed.). New York: Vintage Books.

Corp, N., & Byrne, R. W. (2004). Sex Difference in Chimpanzee Handedness. *American Journal of Physical Anthropology*, *123*(1), 62–68. http://doi.org/10.1002/ajpa.10218.

Courtine, G., Papaxanthis, C., & Schieppati, M. (2006). Coordinated modulation of locomotor muscle synergies constructs straight-ahead and curvilinear walking in humans. *Experimental Brain Research*, *170*(3), 320–35. http://doi.org/10.1007/s00221-005-0215-7.

Dessery, Y., Barbier, F., Gillet, C., & Corbeil, P. (2011). Does lower limb preference influence gait initiation? *Gait and Posture*, *33*(4), 550–555. http://doi.org/10.1016/j.gaitpost.2011.01.008..

Dietz, V. (2002). Do human bipeds use quadrupedal coordination? *Trends in Neurosciences*, *25*(9), 462–467. http://doi.org/10.1016/S0166-2236(02)02229-4.

Duarte, M., Hanna, J., Sanches, E., Liu, Q., & Fragaszy, D. (2012). Kinematics of bipedal locomotion while carrying a load in the arms in bearded capuchin monkeys (Sapajus libidinosus). *Journal of Human Evolution*, *63*(6), 851–858. http://doi.org/10.1016/j.jhevol.2012.10.002.

Elias, L. J., Bryden, M. P., & Bulman-Fleming, M. B. (1998). Footedness is a better predictor than is handedness of emotional lateralization. *Neuropsychologia*, *36*(1), 37–43. http://doi.org/10.1016/S0028-3932(97)00107-3.

Frost, G. T. (1980). Tool behavior and the origins of laterality. *Journal of Human Evolution*, *9*(6), 447–459. http://doi.org/10.1016/0047-2484(80)90002-0.

Giddings, V. L., Beaupre, G. S., Whalen, R. T., & Carter, D. R. (2000). Calcaneal loading during walking and running. *Medicine & Science in Sports & Exercise*, *32*(3), 627–634.

Golomer, E., & Mbongo, F. (2004). Does footedness or hemispheric visual asymmetry influence center of pressure displacements? *Neuroscience Letters*, *367*(2), 148–51. http://doi.org/10.1016/j.neulet.2004.05.106.

Golubitsky, M., Stewart, I., Buono, P., & Collins, J. (1999). Symmetry in locomotor central generators and animal gaits. *Letter to Nature*, *401*(6754), 693–695. http://doi.org/10.1038/44416.

Gundersen, L. A., Valle, D. R., Barr, A. E., Danoff, J. V., Stanhope, S. J., & Snyder-Mackler, L. (1989). Bilateral analysis of the knee and ankle during gait: An examination of the relationship between lateral dominance and symmetry. *Physical Therapy*, *69*(8), 640–650. http://doi.org/10.1093/ptj/69.8.640.

Hall, R. L., & Shereff, M. J. (1993). Anatomy of the calcaneus. *Clinical Orthopaedics and Related Research*, (290), 27–35. Retrieved from http://www.ncbi.nlm.nih.gov/pubmed/8472459.

Hart, S., & Gabbard, C. (1997). Examining the stabilising characteristics of footedness. *Laterality*, *2*(1), 17–26. http://doi.org/10.1080/713754251.

Hayafune, N., Hayafune, Y., & Jacob, H. A. C. (1999). Pressure and force distribution characteristics under the normal foot during the push-off phase in gait. *Foot*, *9*(2), 88–92. http://doi.org/10.1054/foot.1999.0518.

Herzog, W., Nigg, B. M., Read, L. J., & Olsson, E. (1989). Asymmetries in ground reaction force patterns in normal human gait. *Medicine and Science in Sports and Exercise*, *21*(1), 110–4. Retrieved from http://www.ncbi.nlm.nih.gov/pubmed/2927295.

Hesse, S., Reiter, F., Jahnke, M., Dawson, M., Sarkodie-Gyan, T., & Mauritz, K. H. (1997). Asymmetry of gait initiation in hemiparetic stroke subjects. *Archives of Physical Medicine and Rehabilitation*, *78*(7), 719–724. http://doi.org/10.1016/S0003-9993(97)90079-4.

Hopkins, W. D., Phillips, K. A., Bania, A., Calcutt, S. E., Gardner, M., Russell, J., ... Schapiro, S. J. (2011). Hand Preferences for Coordinated Bimanual Actions in 777 Great Apes: Implications for the Evolution of Handedness in Hominins. *Journal of Human Evolution*, *60*(5), 605–611. http://doi.org/10.3899/jrheum.121180.Response.

Hopkins, W. D., Stoinski, T. S., Lukas, K. E., Ross, S. R., & Wesley, M. J. (2003). Comparative assessment of handedness for a coordinated bimanual task in chimpanzees (Pan troglodytes), gorillas (Gorilla gorilla) and orangutans (Pongo pygmaeus). *Journal of Comparative Psychology*, *117*(3), 302–308. http://doi.org/10.1037/0735-7036.117.3.302.

Hutton, W. C., & Dhanendran, M. (1979). A study of the distribution of load under the normal foot during walking. *International Orthopaedics*, *3*(2), 153–157. http://doi.org/10.1007/BF00266887.

Ivanenko, Y. P., Poppele, R. E., & Lacquaniti, F. (2004). Five basic muscle activation patterns account for muscle activity during human locomotion. *Journal of Physiology*, *556*(1), 267–282. http://doi.org/10.1113/jphysiol.2003.057174.

Jones, W. (1870). *A study of handedness*. South Dakota: Capital Supply Company.

Kendall, Florence Peterson, McCreary, Elizabeth Kendall Kendall, H. O. (1983). *Muscles, Testing and Function: Testing and Function.* Lippincott Williams and Wilkins.

Kiriyama, K., Warabi, T., Kato, M., Yoshida, T., & Kokayashi, N. (2005). Medial-lateral balance during stance phase of straight and circular walking of human subjects. *Neuroscience Letters, 388*(2), 91–5. http://doi.org/10.1016/j.neulet.2005.06.041.

Kiyota, T., & Fujiwara, K. (2014). Dominant side in single-leg stance stability during floor oscillations at various frequencies. *Journal of Physiological Anthropology, 33*(1), 25. http://doi.org/10.1186/1880-6805-33-25.

Krasovsky, T., Lamontagne, A., Feldman, A. G., & Levin, M. F. (2014). Effects of walking speed on gait stability and interlimb coordination in younger and older adults. *Gait and Posture, 39*(1), 378–385. http://doi.org/10.1016/j.gaitpost.2013.08.011.

Kuo, A. D. (2001). Energetics of Actively Powered Locomotion Using the Simplest Walking Model. *Journal of Biomechanical Engineering, 124*(1), 113. http://doi.org/10.1115/1.1427703.

Lacquaniti, F., Ivanenko, Y. P., & Zago, M. (2012). Patterned control of human locomotion. *The Journal of Physiology, 590*(10), 2189–99. http://doi.org/10.1113/jphysiol.2011.215137.

Lambrinudi, C. (1932). Use and Abuse of Toes. *Postgraduate Medical Journal, 8*(86), 459–64. Retrieved from http://www.pubmedcentral. nih.gov/ articlerender. fcgi? artid= 2532148&tool= pmcenterz& rendertype=abstract.

Leca, J. B., Gunst, N., & Huffman, M. A. (2010). Principles and levels of laterality in unimanual and bimanual stone handling patterns by Japanese macaques. *Journal of Human Evolution, 58*(2), 155–165. http://doi.org/10.1016/j.jhevol.2009.09.005.

Luetters, C. M., Kelsey, J. L., Keegan, T. H. M., Quesenberry, C. P., & Sidney, S. (2003). Left-handedness as a risk factor for fractures.

Osteoporosis International, *14*(11), 918–922. http://doi.org/10.1007/s00198-003-1450-z.

Lythgo, N., Wilson, C., & Galea, M. (2011). Basic gait and symmetry measures for primary school-aged children and young adults. II: Walking at slow, free and fast speed. *Gait & Posture*, *33*(1), 29–35. http://doi.org/10.1016/j.gaitpost.2010.09.017.

MacLellan, M. J., Ivanenko, Y. P., Massaad, F., Bruijn, S. M., Duysens, J., & Lacquaniti, F. (2014). Muscle activation patterns are bilaterally linked during split-belt treadmill walking in humans. *Journal of Neurophysiology*, *111*(8), 1541–1552. http://doi.org/10.1152/jn.00437.2013.

Matsuda, S., & Demura, S. (2013). Age-related, interindividual, and right/left differences in anterior-posterior foot pressure ratio in preschool children. *Journal of Physiological Anthropology*, *32*(1), 1–7. http://doi.org/10.1186/1880-6805-32-8.

Morris, T., Newby, N. A., Wininger, M., & Craelius, W. (2009). Inter-limb transfer of learned ankle movements. *Experimental Brain Research*, *192*(1), 33–42. http://doi.org/10.1007/s00221-008-1547-x.

Muscolino, J. E. (2008). *The muscle and bone palpation manual with trigger points, referral patterns and stretching* (1st ed.). Elsevier Health Sciences.

Nagano, A., Yoshioka, S., Hay, D. C., Himeno, R., & Fukashiro, S. (2006). Influence of vision and static stretch of the calf muscles on postural sway during quiet standing. *Human Movement Science*, *25*(3), 422–34. http://doi.org/10.1016/j.humov.2005.12.005.

Nelson-Wong, E., Howarth, S., Winter, D. A., & Callaghan, J. P. (2009). Application of autocorrelation and cross-correlation analyses in human movement and rehabilitation research. *The Journal of Orthopaedic and Sports Physical Therapy*, *39*(4), 287–95. http://doi.org/10.2519/jospt.2009.2969.

Neptune, R. R., & Sasaki, K. (2005). Ankle plantar flexor force production is an important determinant of the preferred walk-to-run transition speed. *The Journal of Experimental Biology*, *208*(Pt 5), 799–808. http://doi.org/10.1242/jeb.01435.

Neumann, D. A., Soderberg, G. L., & Cook, T. M. (1988). Comparison of maximal isometric hip abductor muscle torques between hip sides. *Physical Therapy*, *68*(4), 496–502. http://doi.org/10.1093/ptj/68.4.496.

Nissinen, M., Heliövaara, M., Seitsamo, J., & Poussa, M. (1995). Left handedness and risk of thoracic hyperkyphosis in prepubertal schoolchildren. *International Journal of Epidemiology*, *24*(6), 1178–1181. http://doi.org/10.1093/ije/24.6.1178.

Sadeghi, H., Allard, P., Prince, F., & Labelle, H. (2000). Symmetry and limb dominance in able-bodied gait: a review. *Gait & Posture*, *12*(1), 34–45. http://doi.org/10.1016/S0966-6362(00)00070-9.

Sanjaya, K. H., Lee, S., Shimomura, Y., & Katsuura, T. (2014). The influence of laterality on different patterns of asymmetrical foot pressure and muscle activation during gait cycle in manual pushing. *Journal of Human Ergology*, *43*, 77–92.

Sanjaya, K. H., Suherman, Lee, S., Shimomura, Y., & Katsuura, T. (2017). The biomechanics of walking symmetry during gait cycle in various walking condition. In *Proceedings of International Conference on Biomedical Engineering. IBIOMED 2016. IEEE Xplore Digital Library.* (pp. 1–6). http://doi.org/10.1109/IBIOMED.2016.7869813.

Schneiders, A. G., Sullivan, S. J., O'Malley, K. J., Clarke, S. V, Knappstein, S. A., & Taylor, L. J. (2010). A valid and reliable clinical determination of footedness. *PM & R: The Journal of Injury, Function, and Rehabilitation*, *2*(9), 835–41. http://doi.org/10.1016/j.pmrj.2010.06.004.

Seeley, M. K., Umberger, B. R., & Shapiro, R. (2008). A test of the functional asymmetry hypothesis in walking. *Gait and Posture*, *28*(1), 24–28. http://doi.org/10.1016/j.gaitpost.2007.09.006.

Stewart, I., & Golubitsky, M. (2011). *Fearful symmetry: Is God a geometer?* (1st ed.). Cambridge, Mass.: Dover Publications.

Thorstensson, A., Carlson, H., Zomlefer, M. R., & Nilsson, J. (1982). Lumbar back muscle activity in relation to trunk movements during locomotion in man. *Acta Physiologica Scandinavica*, *116*(1), 13–20. http://doi.org/10.1111/j.1748-1716.1982.tb10593.x.

Uomini, N. T. (2009). The prehistory of handedness: archaeological data and comparative ethology. *Journal of Human Evolution*, *57*(4), 411–9. http://doi.org/10.1016/j.jhevol.2009.02.012.

Van Hedel, H. J. A., Biedermann, M., Erni, T., & Dietz, V. (2002). Obstacle avoidance during human walking: Transfer of motor skill from one leg to the other. *Journal of Physiology*, *543*(2), 709–717. http://doi.org/10.1113/jphysiol.2002.018473.

Volpato, V., Macchiarelli, R., Guatelli-Steinberg, D., Fiore, I., Bondioli, L., & Frayer, D. W. (2012). Hand to mouth in a Neandertal: Right-handedness in Regourdou 1. *PLoS ONE*, *7*(8), 3–8. http://doi.org/10.1371/journal.pone.0043949.

Willems, A., Wolters, E., Berendse, H., Lim, I., Rietberg, M., Zijlmans, J., … de Goede, C. (2006). The effect of rhythmic somatosensory cueing on gait in patients with Parkinson's disease. *Journal of the Neurological Sciences*, *248*(1–2), 210–214. http://doi.org/10.1016/j.jns.2006.05.034.

Willems, P. A., Cavagna, G. A., & Heglund, N. C. (1995). External, internal and total work in human locomotion. *The Journal of Experimental Biology*, *198*(Pt 2), 379–93. http://doi.org/10.1007/BF00430237.

Zhao, D., Li, B., & Watanabe, K. (2008). First evidence on foot preference during locomotion in Old World monkeys: A study of quadrupedal and bipedal actions in Sichuan snub-nosed monkeys (Rhinopithecus roxellana). *Primates*, *49*(4), 260–264. http://doi.org/10.1007/s10329-008-0096-z.

Zverev, Y. P. (2009). Spatial parameters of walking gait and footedness. *Annals of Human Biology*. Retrieved from http://www.tandfonline.com/doi/abs/10.1080/03014460500500222.

BIOGRAPHICAL SKETCH

Kadek Heri Sanjaya

Affiliation: Biometrics for Man-machine Interaction Research Group, Research Center for Electrical Power and Mechatronics, Indonesian Institute of Sciences (RCEPM-LIPI)

Education:

Bachelor of Engineering (1996-2002)	Industrial Product Design Faculty of Civil Engineering and Planning Institut Teknologi Sepuluh Nopember (ITS Surabaya)
Master of Engineering (2010-2012)	Humanomics Laboratory Department of Design Science Graduate School of Engineering Chiba University
Doctor of Philosophy (2012-2015)	Humanomics Laboratory Department of Design Science Graduate School of Engineering Chiba University

Business Address:

Research Center for Electrical Power and Mechatronics, Indonesian Institute of Sciences (RCEPM-LIPI), 60th Building, 2nd Floor, Komplek LIPI, Jalan Cisitu No 21/154D, Bandung 40135, Indonesia. Phone: +62-22-2503055, 2504770; Fax: +62-22-2504773

Research and Professional Experience:

No	Year	Research Topic	Funding
1.	2005-2009	Human Factors Engineering in Electric Vehicle	DIPA (Annual Indonesian Government Research Budget)

2.	2008-2009	Human Factors in Manual Materials Handling in a Hospital	Self-funding
3.	2012-2015	Biomechanical Study during Manual Pushing and Locomotion	Monbukagakusho (MEXT) Jepang-part of graduate school scholarship
4.	2016-	Physiological and Biomechanical Study on Driving Fatigue Detection	RisetPro Non-degree scholarship funding
5.	2019-	Walking Symmetry Detecting Smart Footwear as E-Health Monitoring System	Indonesian Ministry for Research, Technology, and Higher Education National Research Incentive Funding (Insinas)

Professional Appointments:

2005-2009	Junior Researcher at RCEPM-LIPI
2006-2009	Principal Investigator for Electric Vehicle Design and Development
2012-2015	Student Member of Japan Society of Physiological Anthropology (JSPA)
2015	Chairman of the 3rd International Conference on Sustainable Energy Engineering and Application (ICSEEA 2015)
2015-	Research Associate at RCEPM-LIPI
2016	Leading RCEPM-LIPI researchers team for a two-month research visit at Coventry University, UK
2016-2018	Thesis advisor for students at Industrial Engineering Department Universitas Katolik Parahyangan (UNPAR)
2017	Chairman of the 5th International Conference on Sustainable Energy Engineering and Application (ICSEEA 2017)
2018	Coordinator of Research Planning, Monitoring, and Evaluation, at RCEPM-LIPI

2018   Internship advisor for student at Industrial Design Department, Institut Teknologi Bandung
2019-   Thesis advisor for student at Master Program, Engineering Physics, Institut Teknologi Bandung

Honors:
2015   Best Presenter at the 3rd International Conference on Creative Industry (ICCI 2015)
2015   Satyalancana Karya Satya for a 10-year dedication as a reseacher at Indonesian Institute of Sciences (LIPI)

Publications from the Last 3 Years:

Publications in Journals

No	Title	Journal	DOI
1.	Review on the application of physiological and biomechanical measurement methods in driving fatigue detection	Journal of Mechatronics, Electrical Power and Vehicular Technology (MEV Journal) Vol 7(1), 2016	DOI: 10.14203/j.mev.2016.v7.35-48
2.	The Effects of Different Bathing Methods on Melatonin Response and a Subjective Evaluation	Journal of Physiotherapy and Physical Rehabilitation, 2016 vol 1(2), 1000112	DOI: 10.4172/2573-0312.1000112
3.	Simulation of lumbar and neck angle flexion while ingress of paratransit (angkot) in Indonesia as a preliminary design study	Journal of Mechatronics, Electrical Power and Vehicular Technology (MEV Journal) Vol 8(2), 2017	DOI: 10.14203/j.mev.2017.v8.70-75

| 4. | Preliminary investigation of sleep-related driving fatigue experiment in Indonesia | Journal of Mechatronics, Electrical Power and Vehicular Technology (MEV Journal) Vol 9(1), 2018 | DOI: 10.14203/j.mev.2018.v9.8-16 |

Publications in Proceeding

No	Conference/Proceeding	Title	DOI
1.	International Conference on Engineering, Science, and Nanotechnology (ICESNANO 2016)	Conceptual framework on the application of biomechanical measurement methods in driving behavior study	DOI: 10.1063/1.4968390
2.	International Conference on Engineering, Science, and Nanotechnology (ICESNANO 2016)	Camouflage design and head measurement characteristic of Indonesian armoured vehicle helmet	DOI: 10.1063/1.4968328
3.	International Conference on Biomedical Engineering (IBIOMED 2016)	The biomechanics of walking symmetry during gait cycle in various walking condition	DOI: 10.1109/IBIOMED.2016.7869813
4.	The 2nd International Conference on Automation, Cognitive Science, Optics, Micro Electro-Mechanical System, and Information Technology (ICACOMIT 2017)	The Mental Chronometry during Simulated Driving Tasks in Various Conditions of Indonesian Subjects	DOI: 10.1109/ICACOMIT.2017.8253376
5.	The 5th International Conference on Sustainable Energy Engineering and Application (ICSEEA 2017)	Ergonomic Assessment on Charging Station Touch Screen Based on Task Performance Measurement	DOI: 10.1109/ICSEEA.2017.8267680

6.	*The 6th International Conference on Sustainable Energy Engineering and Application (ICSEEA 2018)*	*The Comparison between Pre- and Post-Lunch Driving Performance among Morning Person Subjects during a Simulated Driving*	*DOI: 10.1109/ICSEEA.2018.8627137*
7.	*The 6th International Conference on Sustainable Energy Engineering and Application (ICSEEA 2018)*	*A Preliminary Design of Electric Scooter for Sustainable Tourist Transportation*	*DOI: 10.1109/ICSEEA.2018.8627126*

In: A Closer Look at Biomechanics ISBN: 978-1-53615-866-3
Editor: Daniela Furst © 2019 Nova Science Publishers, Inc.

Chapter 5

GEOMETRY AND INERTIA OF THE HUMAN BODY AND THEIR SPORT APPLICATIONS

Wlodzimierz Stefan Erdmann[*]*, PhD, DSc*
Department of Biomechanics and Sport Engineering,
Gdansk University of Physical Education and Sport,
Gdansk, Poland, EU

ABSTRACT

Human biomechanics encompasses body build, forces acting on it, and the results of this action, such as strain, deformation, and movement. Human body morphology can be described from the points of view of body structure, biomaterials, body construction, geometry, and inertia. There are many approaches for obtaining human body geometry and inertia. Scientists have used cadavers, body models, and living people. For external and internal geometry data, the following methods are utilized: direct measurements of the human body (anthropometry), photogrammetry, computerized tomography, and nuclear magnetic resonance. Inertial data are obtained most often by a) densitometry of tissues, body parts, or the whole body; b) mechanical or electronic scales

---

[*] Corresponding Author's E-mail: wlodzimierz.erdmann@awf.gda.pl.

for body mass; c) mechanical lever, photogrammetry for localization of body center of mass; or d) quick release, turntable, calculation methods for moment of inertia. There are several applications within athletic activity for human body geometric and inertial data. Athletic talent identification must take into account the dimensions of the body and their possible development within the course of the athlete's adolescence and youth. Athletic equipment must be built according to body dimensions of the members of current society. Many athletic disciplines are divided onto categories based on body mass (e.g., weight lifting, fighting sports). Localization of the body's center of mass helps in analyses of sport technique, while information on moment of inertia helps in explaining body angular movements.

Keywords: body morphology, geometry, inertia, sport

1. INTRODUCTION

Analysing a human body and its movement is an area of research in biomechanics, kinesiology, ergonomics, medicine (especially orthopaedics), and biomedical engineering. There are several areas of interest within these scientific disciplines including body morphology and the mechanics of tissues, organs, and systems. There is also the mechanics of equilibrium and movement and the influence of environment. Specific individuals are usually investigated, including ordinary humans, workers, passengers, athletes, or patients, but groups of people have also been considered. It is difficult to investigate the human body and there are many different approaches to such investigations. Scientists have used cadavers, body models, and living people (Erdmann 2015).

The biomechanics of sports encompasses several important areas, including talent identification, body morphology, training, body fitness, technique, tactics, interaction of body and devices, influence of environment, control, cooperation of competitors, diagnosis, competition, refereeing, and biological renewal. Geometry and inertia are applied within talent identification, technique analyses, and the interaction of the body and devices (Erdmann 2017).

Body morphology consists of a) body structure—the number of links, degrees of freedom, that is, possibilities of movement in joints, b) biomaterials—natural and artificial, c) body construction—for the purpose of analysis of manipulation, pedipulation, locomotion, or body protection, d) geometry—in three dimensions plus angular measurements, and e) inertia—resistance to external forces, substitution by center of mass (Erdmann 1995, 1997).

It is difficult to analyze the human body for both external and internal reasons. The human body has an irregular geometric shape, so during analyses it is sometimes necessary to substitute simplified geometric models for the actual human body. Body parts are often presented as regular geometric shapes (Hanavan 1969). They are considered in some analyses as rigid solids with no movement of internal tissues while in other, more detailed analyses, they are treated as flexible objects.

Body inertial data include body mass, relative body mass (mass according to volume, i.e., density), radius of center of mass (i.e., location of center of mass according to the acquired reference system), moment of inertia (taking into account body mass and its distribution around the axis of rotation), and radius of moment of inertia. Here, too, irregular body shape does not simplify calculation of these quantities (Erdmann 1999).

Adaptive sports need to take into account different orthoses, prostheses, and other specific devices that give athletes with disabilities a chance to participate in competitions. These devices have different dimensions and inertial quantities. To have the same opportunity to win within the athletic competition, these devices need to have similar biomechanical characteristics.

This chapter describes the geometry and inertia of the human body, including definitions, methods and instrumentation for obtaining data, as well as examples of applications in sports.

2. Geometry of the Body

2.1. A Problem

The geometry of the body describes the space where the body is situated. It encompasses linear, planar, and spatial dimensions. The linear dimensions include straight-linear dimensions and curvilinear dimensions. Planar dimensions include surfaces and planar angles, and spatial dimensions include spaces and spatial angles.

In the history of the geometry of the human body, sculptures by Greek artists demonstrated good body part proportions. Later, the Roman architect Marcus Vitruvius Pollio (known as Vitruvius, lived in the 1st century BCE) was also interested in the proportions of the human body. He presented a model known as the "Vitruvian Man" that was later drawn by Leonardo da Vinci in 1490. In this drawing, Leonardo presented the proportions of the human body based on a square and circle.

The linear dimensions of the human body are usually used for parts of the extremities, while the trunk can be presented as a curvilinear body set of segments. Straight-linear dimensions are used mostly in anthropometry to describe dimensions of the human population (Snyder et al. 1977), in engineering design for construction of furniture or transport vehicles (King and Mertz 1973), and in sport analyses as when comparing step length to the length of the lower extremities (Erdmann 2007). Curvilinear dimensions are used in physiotherapy, for example, to describe the trunk's curves: to the front – *lordosis*; to the back – *kyphosis*; and to the side – *scoliosis*. Planar dimensions can be analysed as a whole body surface (skin surface) or an area perpendicular to the direction of movement, taking into account for calculations of drag, such as air or water resistance. Planar angles are used, for example, to measure angular movement in one-axis joints and when angles between trunk parts are investigated during body posture analyses (Singla and Veqar 2014). The spatial dimensions of the body are analysed when buoyancy is investigated, usually when the body is immersed in water (Donoghue and Minnigerode 1977), while the volume of body parts is taken into account when general research is carried out on the human body, such as when calculating the density of body parts (Clauser et al. 1969). Spatial angles are used during different kinds of movement in two- or three-axes joints.

Data are also gathered on the shape of the body. This is important in the calculation of drag force. Air or water resistance depends on the density of the medium (fluid) in which a movement takes place, the relative velocity of movement, the surface area of a body perpendicular to the direction of movement and the shape of the body. The latter is given in a value of coefficient /1/:

$$F.d = 1/2\rho \times v \times A \times c.d \qquad (1)$$

where $F.d$ is drag force, ρ is mass density of the fluid, v is the flow velocity of the object relative to the fluid, A is the surface area perpendicular to the direction of movement, and $c.d$ is the drag coefficient.

A regular prism, for example, has a shape coefficient of 1.0, while the shape of a falling drop has a coefficient of 0.05 (Figure 1).

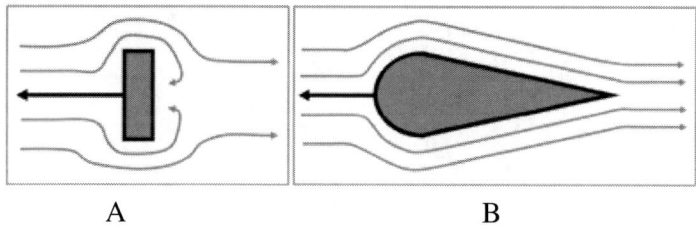

Figure 1. Moving of objects with different shapes: A – regular prism (turbulent flow); B – falling drop (laminar flow).

2.2. Obtaining External Body Geometry

The linear dimensions of the whole body or of its parts are measured in different ways, including the following:

1) With the help of an anthropometer, that is, a narrow calibrated tube with a horizontal arm that can move along and perpendicularly to a tube. The tube is based on a floor in the center and behind the feet, held vertically, and the end of the arm touches a specific body landmark (Figure 2A).
2) Using an image (a photograph, a movie frame, or video) where an object of known dimension is seen near the investigated body, which helps to calculate the scale of the image. Images are usually taken from two (front and side), three (front, side, and back), or four (front, side, back, and top) directions (Figure 2B).
3) Introducing regular geometric shapes to body parts with cross-section shape that forms an ellipse or other planar geometric figure (Figure 2C).
4) Using a three-dimensional laser scanner to obtain linear, planar, and spatial dimensions of the body and its segments (Figure 2D).

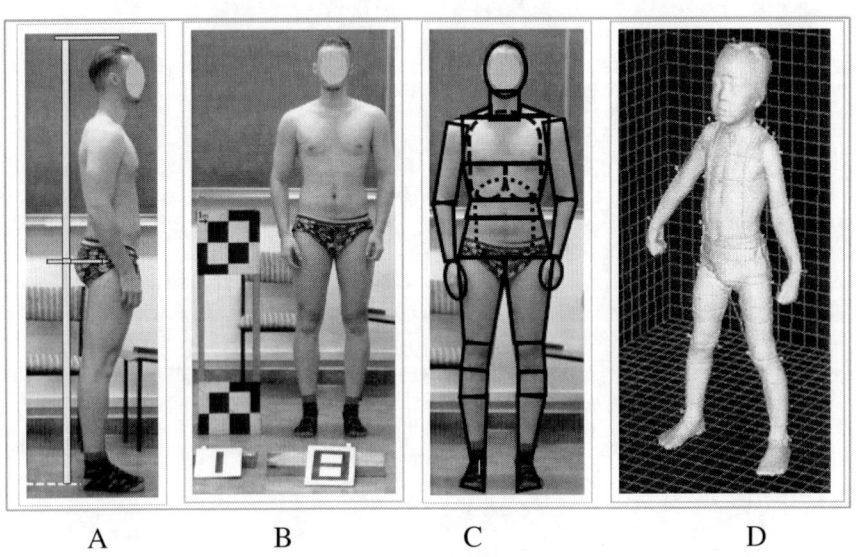

Figure 2. Different approaches to obtain external body dimensions using: A – anthropometer; B – body image from a photograph; C – regular geometric shapes (dashed line – thorax, i.e., lungs, ribs; dotted line – internal tissues of abdomen); D – image obtained from a laser scanner (courtesy: Hamamatsu).

2.3. Obtaining of Internal Body Geometry

Up to the invention of x-rays by W. C. Roentgen (Nobel prize of 1901), internal images of the body were obtainable only by cutting cadavers. The next significant breakthroughs for internal body imaging were computer-assisted tomography (CAT) by G. N. Hounsfield and A. M. Cormack (1979 Nobel Prize) and magnetic resonance imaging (MRI) by P. Mansfield and P. Lauterbur (2003 Nobel Prize).

Using CAT, the author of this chapter obtained internal body images to have geometric dimensions of the anatomic organs of the male trunk (Figure 3). These images helped in the division of the trunk onto parts. For example, the upper part of the trunk consists of the thorax and shoulder girdle (Figure 4), the upper mid part of the trunk consists of the abdomen and thorax, while the lower mid part consists of the abdomen and pelvis. To calculate the mass of a layer or its fragment that belongs to particular body part, data on the density and volume of the tissues are needed. Volume was obtained by multiplication of the layer's height and the area of the surface of the particular tissue (Figure 5).

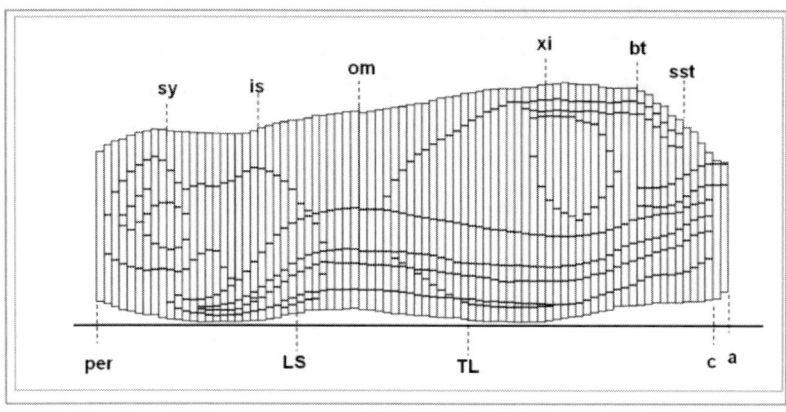

Figure 3. Image obtained from computerized tomography, i.e., a set of layers of 8 mm width between the anthropological landmarks *acromion (a)* and *perineale (per)* (Erdmann 1995, 1997).

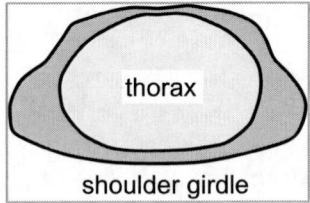

Figure 4. Differentiation of particular body parts in the image of a layer (Erdmann 1995, 1997).

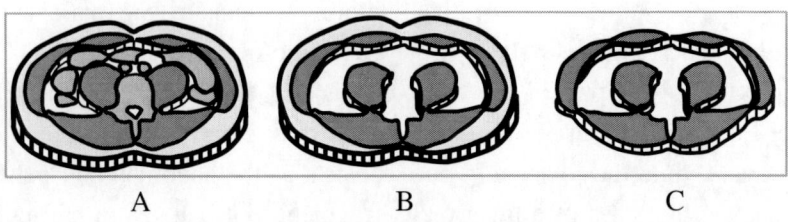

Figure 5. Tissues within a layer at the level of umbilicus: A – bone, digestive, muscle, and adipose (fat); B – muscle and adipose; C – muscle (Erdmann 1995, 1997).

2.4. Body Geometry and Sport

The most common information available about body dimensions is the height of athletes. Everybody knows gymnasts are small and basketball players are tall. There is, however, much larger knowledge on body geometry within sports. In many athletic disciplines, not just the overall height of the body is important but also the lengths of body parts. For example, in basketball the height of the arm joint from the floor and the length of the upper extremities are both important. In athletics, the length of the lower extremities of a runner is important, and an index of stride length according to the length of the legs can be calculated.

According to Usain Bolt, his coach Glen Mills opposed his running at 100 m (Bolt and Allen 2013). He worried that Bolt's tall body would be a disadvantage during the starting portion of the run. Despite Mills' interest in biomechanics (Levy and Mills 2009), he did not perform a deep biomechanical analysis of the function of human body during sprint running. A long trunk results in long internal muscles such as *iliopsoas*. This muscle is the main propeller of the lower extremity in the hip joint during flexion movement. The longer the muscle, the faster the movement of its insertion point during muscle contraction, and the faster the movement of the extremity. A longer bone (femur) also creates a longer arm for the lever, which gives a higher value for the moment of force acting on a bone (Erdmann 2016). Similarly, Michael Phelps, having longer than normal upper extremities, had an advantage in the propulsion of his body in water, where much of the thrust in swimming comes from arm propulsion (Hadhazy and Weiner 2008).

It is possible to adjust athletic devices for different competitor body dimensions. For example, in volleyball the net is set up at different heights (Figure 6A), in artistic gymnastics, the parallel bars are adjusted to the width of the athlete's body (Figure 6B), while in rowing a stretcher can be moved forward or backward according to the length of rower's lower extremities (Figure 6C).

From many years disabled athletes could participate in competitions specially organized for them using specially designed prostheses (Figure 7A and B). This kind of competition developed in the Paralympics movement. Recently, new area of sport for disabled has emerged call the Cybathlon, which is a competition for severely disabled people using technology to assist them substantially in their movement. The first Cybathlon was organized in Zurich in 2017 by Robert Riener (Riener 2017), which included, in addition to different kinds of computerized prostheses for the limbs, orthoses to support the body during posture and movement. One of such device is the exoskeleton – Figure 7C.

The role of prostheses is to substitute for a lost extremity or its part. Prostheses must be of proper dimensions. The length of a prosthesis plus the healthy remainder of the extremity from one side must be equal to the other full healthy extremity. An exoskeleton must properly fit the individual body.

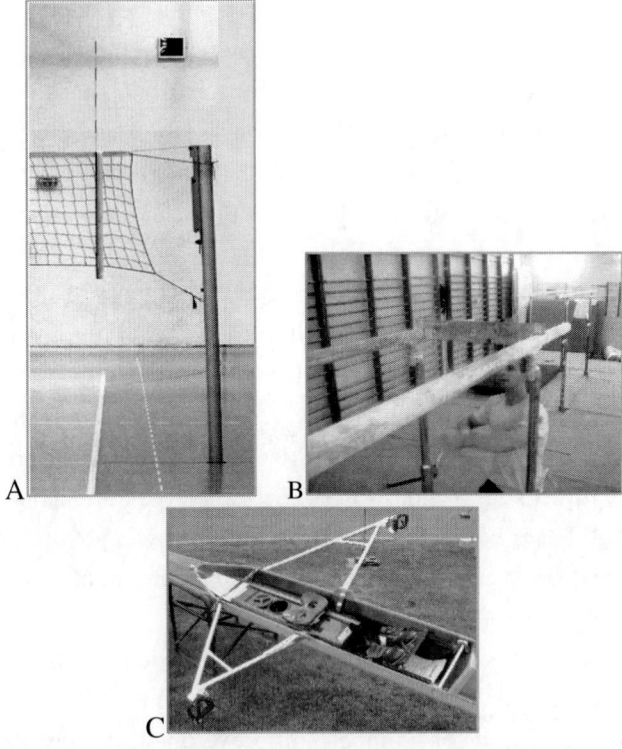

Figure 6. Athletic devices adjusted to athletes' body dimensions in: A – volleyball; B – artistic gymnastics; C – rowing.

Prostheses are designed to restore a lost function, but for the runners they should not give advantage in stride length. The International Paralympics Committee (IPC) uses a formula based on the measurement of the athlete's body and upper extremities' span. This dimension plus 3.5% is used to estimate the athlete's height including the length of the prostheses. This encompasses variations in the able-bodied runners who have long lower extremities and the fact that they are taller when running on their toes.

During the Paralympic Games in London in 2012, a South African runner claimed he was unfairly beaten by another runner (from Brazil, who won the gold medal in the 200 m final) saying that the Brazilian used longer prostheses, allowing him to have longer strides (Lees 2012). The South African runner, however, had longer stumps, so his blades were shorter and the Brazilian had shorter stumps, so it appeared that his prostheses were too long, but according to the IPC, the Brazilian's blades were legal (Figure 8).

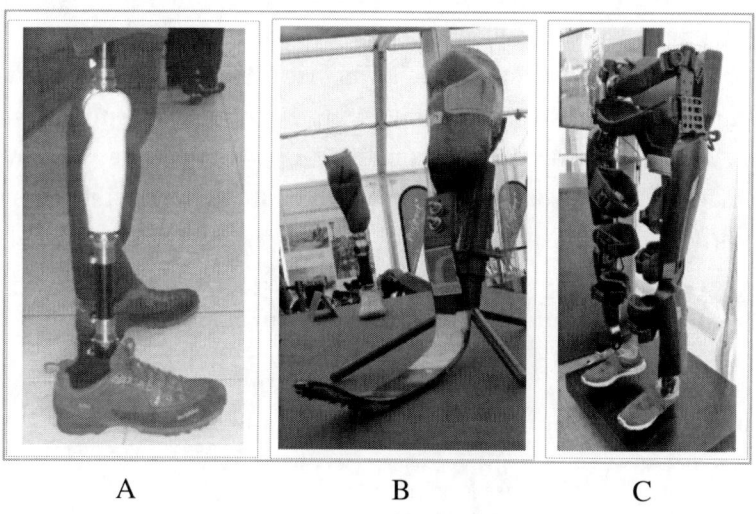

Figure 7. Biomedical engineering approach to the geometry of the human body: A and B – prostheses must be equal in length to the healthy extremity; C – patient's body must fit properly into the exoskeleton.

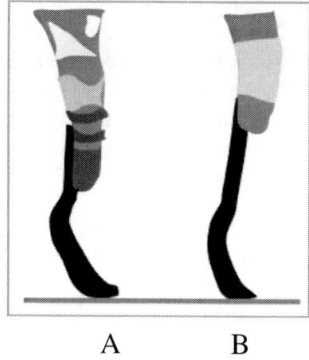

Figure 8. Different lengths of stumps and blades for South African (A) and Brazilian (B) runners; overall length of the stump plus prosthesis was similar for two runners during the 200 m run at the 2012 Paralympic Games in London.

In several sports, the surface area perpendicular to the direction of movement is diminished as much as possible. This can be seen in cycling, alpine skiing, luges, and speed skating. Together with other quantities described earlier, this surface area influences drag (Figure 9). In other situations, athletes seek to enlarge drag, particularly in situations where the wind is from the back of the body (i.e., a tailwind). Speed records for the 100 m can be acknowledged when the tailwind has velocity of no more than 2.0 m/s. Another situation is in vertical movement, as when parachuting where athletes free fall with extended extremities. In ski jumping, competitors act in the horizontal direction like javelins where drag should be minimized and in vertical direction they act like parachute to maximize drag.

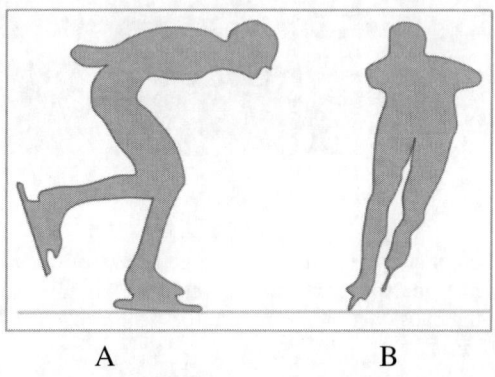

A   B

Figure 9. Silhouettes of a leaning speed skater to diminish drag, showing surfaces in sagittal plane (A) vulnerable to side wind in open-air facilities and in frontal plane (B) where drag associated with the front movement has to be overcome.

The volume of internal organs such as the heart and lungs has an important influence on athletes' achievements. Cyclists and cross-country skiers tend to have the largest heart volumes in terms of pumping a large amount of blood (and large volume of oxygen [VO_2]) in a given time. The highest value achieved for VO_2 max was almost 100 ml/kg/min (Wood 2001). The American swimmer Michael Phelps, the best swimmer in the history of the sport, has perhaps the largest lung volume, at 12 dm^3, which is twice the average volume (Hanlon and Smith 2012).

The volume of body parts and the body as a whole differs substantially in different athletic disciplines. This is especially seen when comparing disciplines in which athletes act for a long time (e.g., in long-distance running; Figure10A) and those in which they act for a shorter time but have a leaning-trunk body configuration such as in speed skating (Figure 10B). The highest volume for body parts are found in athletes in the disciplines of weightlifting, body building, and shot putting.

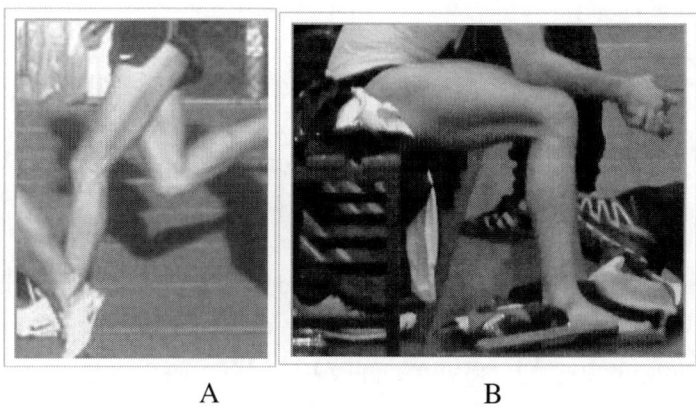

A　　　　　　　　B

Figure 10. Significant difference in the thigh volume of a long distance runner (A, during the run) and speed skater (B, at rest after the run), while the volumes of their shanks are similar.

The volume of the whole body is investigated when the density of the body is being checked. The whole body is immersed into water in a tank, and the water that had been displaced is a measure of the body's volume.

3. INERTIA OF THE BODY

3.1. A Problem

Inertia is the resistance of the body when the state of the body is to be changed, such as when the body is moved, accelerates, or when the direction of movement is changed, or the body decelerates, or stops. In all of these situations a force must be used to change the state of the body; this can be own force (i.e., muscle strength) or external forces.

There are two measures of inertia:

1) In transversal movement, body mass *m* is a measure of inertia. It depends on the density and volume of the object:

$$m = \rho \times V \tag{2a}$$

$$m = \int (\rho V)\, dV \tag{2b}$$

where: *m* is mass, ρ is density, and *V* is volume.

2) In rotational movement mass and its location according to the axis of rotation is a measure of inertia (called the mass moment of inertia *I* or, in short, moment of inertia). There are different approaches to the calculation of moment of inertia, which will be presented below.

The density of a body depends on the kind of material from which the body is composed. It can be calculated by conversion of equation (2a):

$$\rho = m / V \tag{2c}$$

An important element of the body is its center of mass. This is an agreed point not attached to the material representation of the body, that is, it can move out of the body as in the example of a circle. Center of mass represents a location in space where the whole mass of the body can be concentrated. The vector of gravity force or inertial force is attached to this point. The radius of center of mass is the distance from the center of mass to the acquired reference system (in one, two, or three dimensions). This location can change over time. Moment of mass (static moment) is a product of the mass and the radius of the center of mass:

$$Mm = m \times d \tag{3}$$

where *Mm* is moment of mass, *m* is mass, and *d* is the radius of the center of mass.

The simplest moment of inertia is that of a point-sized body with mass *m* and at distance *d* from the axis of rotation *N* (mathematical pendulum; Figure 11A):

$$I_N = m \times d^2 \qquad (4)$$

where I_N is the moment of inertia according to the axis *N*, *N* is the axis of rotation going beyond the center of mass, *m* is mass, and *d* is distance of the body to the axis of rotation *N*.

When a body in the shape of a solid moves around the axis *C* that goes through its center of mass, the moment of inertia equals (Figure 11B):

$$I_C = \Sigma \, (m.i \times r.i^2) \qquad (5)$$

where: I_C is the central moment of inertia according to the axis *C*, *C* is the axis of rotation going through the center of mass, *m.i* is the mass of the *i*-th particle of the body, and *r.i* is distance of the particle of the body to the axis of rotation *C*.

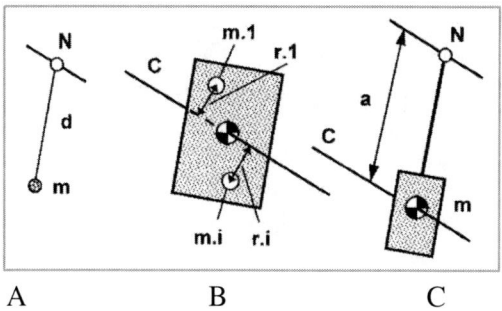

Figure 11. Schemes presenting bodies rotating around axes *N* (A and C) and *C* (B) to calculate their moment of inertia. Black and white circles show the location of the center of mass (Erdmann 2015).

When a solid moves around the axis N that goes beyond its center of mass, the moment of inertia equals (Figure 11C):

$$I_N = I_C + (m \times a^2) \tag{6}$$

where: I_N is the moment of inertia according to the axis N, N is the axis of rotation going beyond the center of mass (non-central axis), I_C is the central moment of inertia according to the axis C, m is the mass of the body, and a is the distance between axes N and C. This is the parallel axes theorem.

3.2. Obtaining the Mass of the Body

The mass of the human body changes from about 3–5 kg of newborn boys up to over 600 kg for extremely obese adult people. Measurement of body mass is usually performed using a scale. There are different types of scales, some of which are shown in Figure 12.

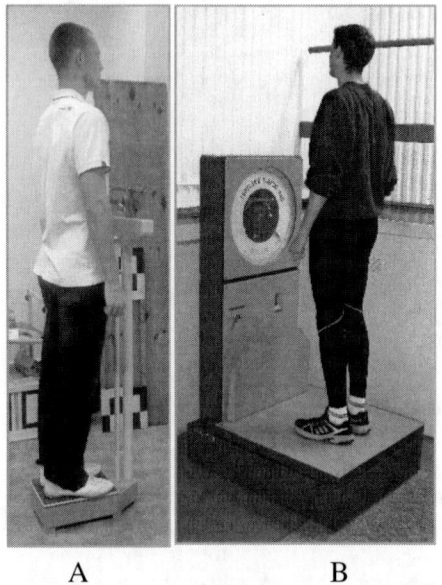

Figure 12. Obtaining the mass of the whole body using a lever scale (A) or spring scale (B). When a movement takes place in clothes, the mass of the body with clothes should be measured.

The mass of body parts can be obtained in different ways. In the nineteenth century, a dead body could be frozen and cut onto segments, and the mass of the segments was measured (Harless 1857). This approach was used by several researchers. In the twentieth century, an important investigation was performed by Hanavan (1964), who presented human body parts that resembled mathematical solids. Using the density of body segments, he could then calculate their mass (Figure 13A).

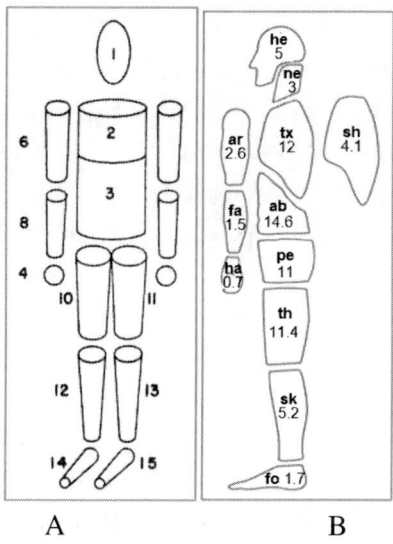

Figure 13. Two different approaches for obtaining inertial values of body segments: A –using mathematical solids and their density (Hanavan 1964); B –using personal data on the volume of body parts of subjects and density; here data of young, fit males are presented, as obtained by Kowalczyk (2013).

Clauser et al. (1969) measured the geometry of body parts (length, breadth, and circumference) and compared this with the inertial values of those parts. Based on this, they then presented regression equations to calculate personal inertial quantities (mass and location of center of mass) for all body parts. Unfortunately, the trunk was treated as a one segment only. The division of the trunk onto four anatomical segments was used by Dempster (1955), who presented mean data for all body parts from eight cadavers. Personalized data for obtaining inertial quantities of all body parts were presented by Zatsiorsky and Seluyanov (1979), who used gamma rays to divide the trunk onto three segments by planes perpendicular to the longitudinal axis of the trunk.

Further methods for acquiring inertial quantity values include computerized tomography (CT) and magnetic resonance images (MRI), which use x-rays and radio frequency energy, respectively. Erdmann (1995, 1997) presented a method using data on the density of tissues (Erdmann and Gos 1990) and the volume of tissues obtained with CT (see Figures 3–5). Using data on density and volume he then calculated the mass of tissues and the mass of the five trunk parts. He also gave a procedure for calculating personal inertial values for the human trunk by measuring the volume of the body parts of investigated person and using density data. This method was used by Kowalczyk (2013) when he investigated a group of high jumpers and a reference group of young, fit males (Figure 13B). The very good validation of both methods (from Clauser and Erdmann) was confirmed by Erdmann and Kowalczyk (2015).

3.3. Localization of Center of Mass and Obtaining Its Radius

Localization of the center of mass can be done directly or indirectly. The direct method uses a board lying on two supports (Fig. 14 A). Here:

$$D = (W \times L) / Q \tag{7}$$

where: D is the distance from the board's end to the center of mass, W is the value shown by the scale, L is the length of a board, and Q is the weight of the body.

Before further calculations, the weight of the board should first be subtracted from the value shown by the scale. The location of the center of mass is described according to the length/height of the whole body (Figure 14 B). This value usually equals 53%–55% of the body height (Bober 1965), but among athletes who have extensively developed the upper body and also have slim lower extremities, this value can even reach slightly above 60%.

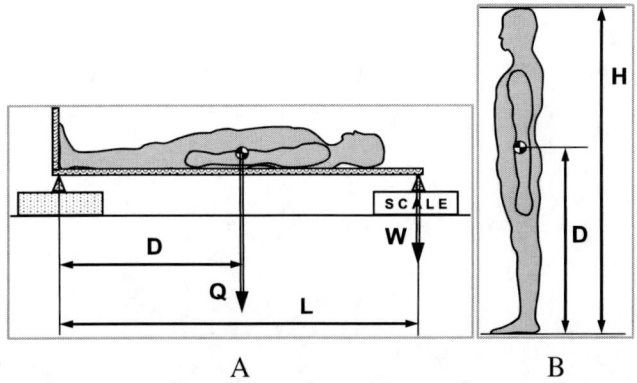

Figure 14. Directly obtaining the location of the center of mass: A – using a board (L – length of a board, Q – weight of the body, W – value shown by the scale, D – distance from the board's end to the center of mass); B – location of center of mass according to height H (Erdmann 2015).

Indirect localization of the whole body center of mass can be performed using an image of the body. Here masses and localization of the body parts' centers of mass (radii of the centers of mass) are needed (Figure 15). Two methods can be used: 1) summing of masses (graphic) or 2) summing of moments of masses (analytic). In the first method, the position of the common center of mass (distance a) can be found using the equations /8a/ and /8b/ (Figure 16):

$$m1 / (m1 + m2) = a / (a + b) = m1 / (m1 + m2) = a / c \qquad (8a)$$

$$a = [(m1 / (m1 + m2)] \times c \qquad (8b)$$

where: $m1$ is the mass of a smaller body, $m2$ is the mass of a larger body, a is the distance from the larger body to the common center of mass, b is the distance from the smaller body to the common center of mass, and c is the distance between the centers of mass of the two bodies.

Here is an example of finding the common center of mass: mass of hand (ha) and forearm (fa) are added to form the mass of the koarm (ko) – Figure 17A and B. Then mass of arm (ar) is added to form the free upper extremity (fu). Next, the mass of the shoulder (sh) is added to form the mass of the (full) upper extremity (ue). By adding the masses of other the parts, one can obtain the location of the center of mass of the whole body (wb).

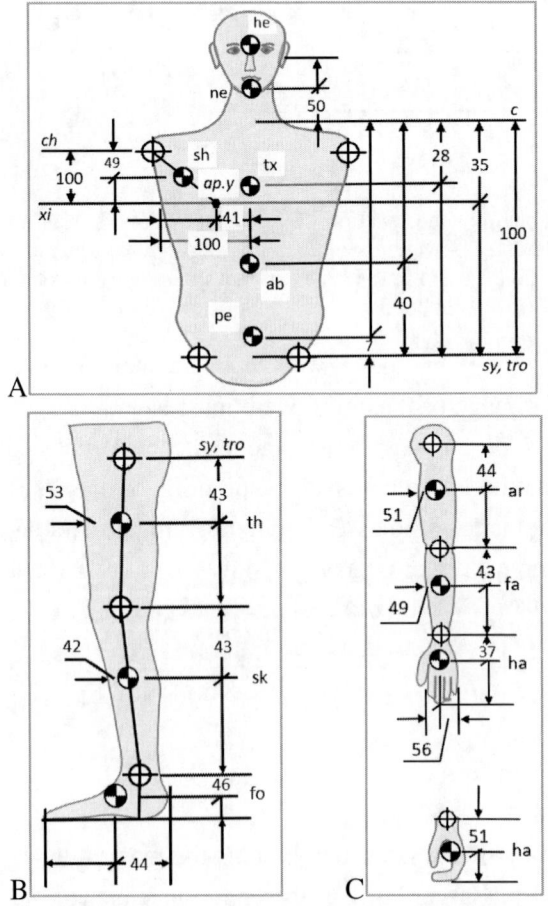

Figure 15. Location of body parts' centers of mass for: A – centroid (head, neck, and trunk); B – lower extremity, C – upper extremity (Clauser et al. 1969, Erdmann 1995, 1997).

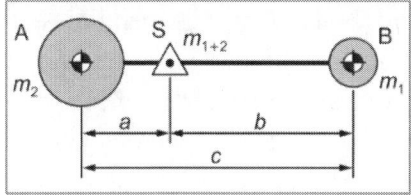

Figure 16. Obtaining the common center of mass of two different bodies whose mass and center of mass positions are known.

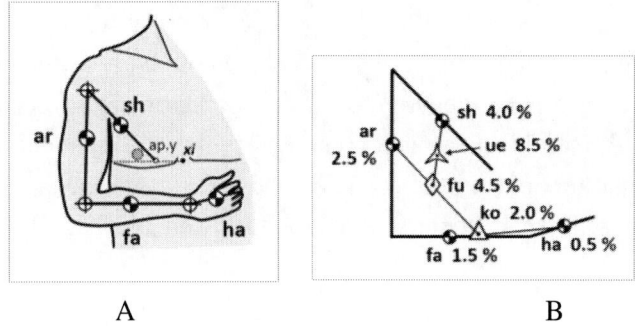

Figure 17. Obtaining the common center of mass of the upper extremity using the graphical method (Erdmann 2000): A – a scheme of the upper extremity and centers of mass of the extremity's parts; B – obtaining the common centers of mass of the hand (ha) + forearm (fa) = koarm (ko), koarm + arm (ar) = free upper extremity (fu); free upper extremity + shoulder (sh) = upper extremity (ue); mass of body parts in percentages of the mass of the whole body. They has been taken from Fig. 13B and rounded to the nearest 0.5%.

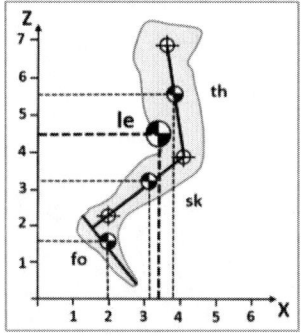

Figure 18. Obtaining the common center of mass of the lower extremity using the analytical method; th – thigh, sk – shank, fo – foot, le – lower extremity (Erdmann 2000).

Summing of the moments of masses method uses a reference system (two axes for planar analysis) drawn near the body image (Figure 18). Then the masses are projected onto the reference axes. The distance of the center of mass (radius of mass) is multiplied by the mass of the body part, yielding the moment of mass. The moments of all body parts according to both axes are obtained. When the sum of the moments is divided by the mass of the whole body, then the principal radius of mass is obtained. The center of mass of the whole body lies at the crossing of both principal radii (for both axes). Figure 18 shows the location of the center of mass of the lower extremity, but this can be done for the whole body or the body with equipment as well. This method is used during automated procedures and computerization of localization of center of mass. A scanner finds the location of markers put on the joints of body parts, and then the software localizes the segments' centers of mass, followed by that of the whole body center of mass.

3.4. Obtaining the Moment of Inertia

Obtaining the moment of inertia means obtaining the resistance of the body, that is, its mass and the distance of the mass from the center of rotation. This quantity can be obtained using a pendulum, a quick release method, or a turntable using an accelerometer, or the calculation method.

For the pendulum, the following equation is used /9/:

$$I = (D \times T^2) / (4 \pi^2) \tag{9}$$

where: I is the moment of inertia, D is the moment of force acting on the pendulum, and T is the period.

For application of an accelerometer, the following equation is used /10/:

$$I = dM / d\alpha \tag{10}$$

where: *I* is the moment of inertia, *dM* is the instant moment of force, and *dα* is the instant angular acceleration.

The moment of inertia of the lower extremity in relation to the frontal hip axis can be calculated by summing up the moments of inertia of its parts (Erdmann 1999):

$$I.hip.LE = I.hip.th + I.hip.sk + I.hip.fo \qquad (11)$$

where: *I.hip* is the moment of inertia according to the hip axis, *LE* is the lower extremity, *th* is the thigh, *sk* is the shank, and *fo* is the foot.

Moments of inertia according to the frontal hip axis of the above parts using equation /6/ are as follows:

$$I.hip.th = I_C th + (m.th \times a.th^2) \qquad (12a)$$

$$I.hip.sk = I_C sk + (m.sk \times a.sk^2) \qquad (12b)$$

$$I.hip.fo = I_C fo + (m.fo \times a.fo^2) \qquad (12c)$$

The radius of gyration (gyradius) *R* is the distance between the axis of rotation and the point where the whole mass of the body is assumed to be concentrated. The moment of inertia about the given axis would be the same as its actual distribution of mass. One can obtain 1) the central radius of gyration R_C and 2) the non-central radius of gyration *R* as follows:

$$R_C = \sqrt{I_C / m} \qquad (13)$$

$$R = \sqrt{I / m} \qquad (14)$$

Zatsiorsky et al. (1981) presented the central radii of gyration for parts of the lower extremity according to the frontal axis. They are as follows: for the thigh, 0.267 of the thigh length; for the shank, 0.275 of the shank length; and for the foot, 0.245 of the foot length. Using the central radius of gyration and the length of a body part, one can calculate the central moment of inertia for particular parts of the lower extremity:

$$I_C.th = m.th \times R_C.th^2 \qquad (15a)$$

$$I_C.sk = m.sk \times R_C.sk^2 \qquad (15b)$$

$$I_C.fo = m.fo \times R_C.fo^2 \qquad (15c)$$

The calculation method for obtaining the moment of inertia uses a graphical model of the lower extremity. Such a model according to the frontal axis is presented in Figure 19, and equations 12a–12c are used.

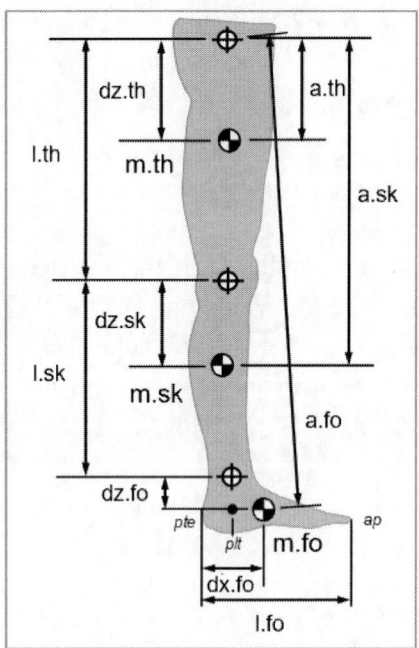

Figure 19. Model of lower extremity showing all necessary quantities for calculation of moment of inertia according to hip axis (Erdmann 2000): th – thigh, sk – shank, fo – foot, m – mass, l – length, d – radius of the center of mass to the nearest joint axis or to the rear edge of the foot, a – radius of the center of mass to the hip axis, z – vertical direction, x – horizontal direction, *ap – acropodion, pte – pternion, plt – processus lateralis tali.*

The fully flexed lower extremity has the value of the moment of inertia according to the frontal hip axis about one third of that for the straightened extremity (Figure 20).

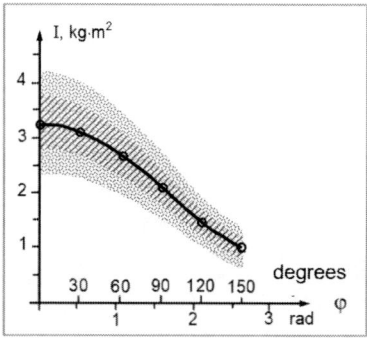

Figure 20. Moment of inertia I of the lower extremity for 100 young, fit men according to knee angle φ: solid line – mean; dashed surface – 1 S.D.; dotted surface – 2 S.D. (Erdmann 1987).

3.5. Body Inertia and Sport

Knowledge of whole body mass is necessary to qualify for certain categories in weightlifting, fighting sports, and rowing. The highest body mass is recorded in sumo, where it can reach over 300 kg. The lightest body mass is found among jockeys in horse riding, the lightest category in boxing, or for coxswains in rowing (i.e., about 50 kg).

The mass of body parts differs substantially. The lightest is the hand (from the wrist to the tip of the longest finger). It is usually less than 1.0% of the whole body mass. The mass of a femur is about 10% for non-trained lean people to about 12% for trained males.

Disabled athletes who ride on a wheelchair have a very small amount of mass at the lower extremities (see Figure 21A and B). A special approach in calculating relative mass of body parts should be taken if the disabled person lacks parts of the body (Figure 21C).

The location of the whole body center of mass is important in assessing the technique of movement among athletes in many different athletic disciplines. It is helpful in many athletic biomechanical analyses. For example, in alpine skiing one can assess movement technique by locating the center of mass according to the skis (forward and backward), according to the ground/snow (up and down), or according to the pole (left and right). In the long jump, center of mass helps in dividing the jumping distance onto fragments (Figure 22A), and in the high jump the location of the center of mass for the erect standing body of the competitor is compared with the location of the center of mass over the bar during the jump (Figure 22B). In judo, an attacker needs to lower his or her body compared to the opponent to execute a throw with higher efficiency, which is assessed by the location of the centers of mass.

Figure 21. Disabled athletes as participants in marathon competition have much smaller lower extremity mass and much higher upper extremity mass than abled people: A – on a wheelchair, with lower extremities bent in front of the trunk; B – on a wheelchair, with lower extremities bent under the trunk; C – on a hand bicycle, with one lower extremity and a stump as the extension of a lying trunk.

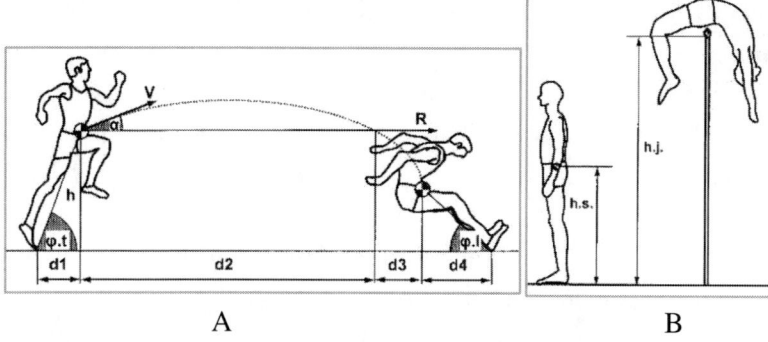

Figure 22. Center of mass used to assess technique during jumps: A – in the long jump to divide the whole distance onto parts; B – in the high jump to assess efficiency of the jump by comparing standing height (h.s) and jumping height (h.j) (Erdmann 2007).

Yet another example of applying the location of the center of mass is in hurdle running. The runner needs to lower his or her body while clearing the obstacle to rise the center of mass as little as possible while not touching the hurdle. The difference in the height of the center of mass while clearing the hurdle and the height of the hurdle makes it possible to assess clearing technique (Figure 23).

Figure 23. Assessment of technique in clearing an obstacle during running 400 m hurdles. Location of competitor's whole body center of mass (h.wb) is compared with the height of the hurdle (h.hu). The difference (Δh) should be as small as possible.

During a pirouette in figure skating, one starts the action with the extremities far apart from the trunk, which creates angular momentum (i.e., the product of the moment of inertia and angular velocity). By drawing in the extremities to the axis of rotation while maintaining the same angular momentum, the moment of inertia is smaller and the angular velocity is higher.

When a judo attacker is going to throw an opponent, he or she at first tries to pull the opponent closer to his or her body. By drawing in the opponent's body towards the attacker's body, the moment of inertia would be smaller, and the angular velocity during the attack would be higher.

During the discus throw, it is better to have the disk far away from the axis of rotation. In this way the discus has a higher linear velocity at release, but having the discus further from the axis of rotation creates a high value for the moment of inertia. Because the movement effect is important, the discus thrower should strengthen his or her muscles to create more opportunity for rotation with high angular velocity and the discus far from his or her shoulder.

Figure 24 shows the two positions of the lower extremity of the sprint runner. One can see that while the lower extremity is straightened, it has a supporting role and the centers of mass of the parts of the extremity are far away from the hip joint axis. When the lower extremity is flexed during recovery, its heel almost touches the buttock. In this way, the moment of inertia of the lower extremity is significantly diminished, so the angular velocity, ω, can be much higher in comparison to the smaller flexion of the extremity.

For many people, including Usain Bolt's coach, it was a surprise when Bolt (body height, 196 cm) obtained a very good time in the 100 m sprint when he was still a young competitor. When a person's body is large, however, his or her lower extremities are usually also large, so the moment of inertia of the lower extremity is high. This is not the case for Usain Bolt: his lower extremities are slim, so the moment of inertia of his lower extremities is small compared to the length of his legs. He can therefore obtain a high angular velocity for his lower extremities, while also taking long strides during the sprint.

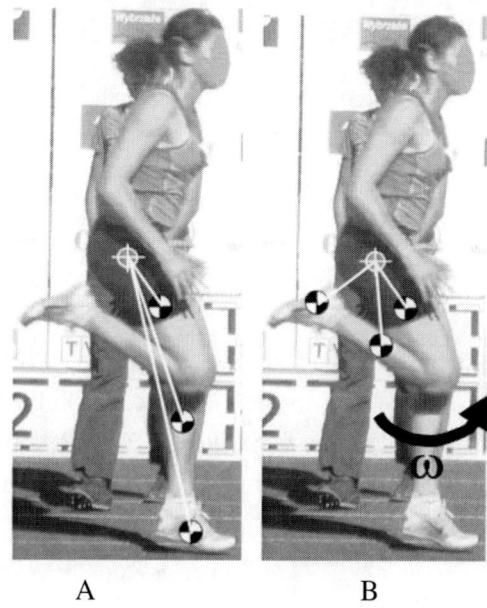

Figure 24. Comparison of distances of the parts of the lower extremity according to the frontal hip axis for straight (A) and flexed (B) lower extremity. During the latter state, a high angular velocity, ω, value can be obtained.

FINAL REMARKS

There are several important geometric and inertial quantities that help to analyze athletic movement. One can here change the assessment from qualitative to quantitative. Some wealthy sports clubs or national teams have their own analysts; as the manager, Alexander Ferguson, has remarked, an analyst is a full member of the team (Ferguson and Moritz 2016). Many other sports clubs do not have such a co-worker, however. Based on the findings presented in this chapter, the present author believes that, at the beginning of the twenty-first century, it is essential to have an analyst to apply knowledge of the athletes' **geometry** and body inertia to the training regime. To use analytical quantities, one should be accustomed to the natural sciences and statistics to present and use data on morphology, biomechanics, physiology, motor learning, and the kinesiology of sports quantitatively, as well as being able to show relationships and dependences.

REFERENCES

Bober T. (1965) Contribution to the research on center of gravity in humans. (in Polish). *Physical Culture (Kultura Fizyczna)*, 3:156-160.
Bolt, U. St. L., Allen, M. (2013) *Usain Bolt. Faster Than Lightning: My Autobiography*. Polish Edition 2014, Wroclaw: Publisher Bukowy Las.
Clauser, C. E., McConville, J. T., Young, J. W. (1969) *Weight, volume, and center of mass of segments of the human body*. Technical Report AMRL-TR-69-70, Aerospace Medical Research Laboratory, Aerospace Medical Division, Air Force Systems Command. Dayton, Ohio: Wright-Patterson Air Force Base.
Dempster, W. T. (1955) *Space requirements of the seated operator. Geometrical, kinematic, and mechanical aspects of the body with special reference to the limbs*. Technical Report WADC-TR-55-159, Wright Air Development Center, Air Research and Development Command, United States Air Force Dayton, Ohio: Wright-Patterson Air Force Base.
Donoghue, E. R., Minnigerode, S. C. (1977) Human body buoyancy: a study of 98 men. *Journal of Forensic Science*. 22:3:573-579.
Erdmann, W. S. (1987) Individual moment of inertia of the lower extremity of the living young adult male. In *Biomechanics X-B, Proceedings of the Xth International Congress of Biomechanics, Umeå, Sweden, 15-20 June 1985*, edited by B. Jonsson, 1049-1053. Champaigne, Ill.: Human Kinetics Publishers.
Erdmann, W. S. (1997) Geometric and inertial data of the trunk in adult males. *Journal of Biomechanics*, 30:7:679-688.
Erdmann, W. S. (1999) Geometry and inertia of the human body: review of research. *Acta of Bioengineering and Biomechanics*, 1:1:23-35.
Erdmann, W. S. (2000) *Biomechanics. A handbook for exercises* (in Polish). Gdansk: May.

Erdmann, W. S. (2007) Morphology biomechanics of track and field competitors. In *XXV International Symposium on Biomechanics in Sports*, edited by H. Menzel, 19-26. Belo Horizonte, MG: State University of Minas Gerais.

Erdmann, W. S. (2015) *Biomechanics. Basics for biomedical engineering major* (in Polish). Gdansk: Gdansk University of Technology.

Erdmann, W. S. (2016) *Gdansk biomechanical report on athletic sprinting 2016*. Report ZBIS, December 14, 2016. Gdansk: Gdansk University of Physical Education and Sport.

Erdmann W. S. (2017) Geometry and inertia of the body as important factors in sport biomechanics. *Visnik chernigivskogo natsionalnogo pedagogichnogo universitetu*, Chernigiv (Ukraine), 147:1:94-100.

Erdmann, W. S., Gos, T. (1990) Density of trunk tissues of young and medium age people. *Journal of Biomechanics*, 23:9:945-947.

Erdmann, W. S., Kowalczyk, R. (2015) A personalized method for estimating center of mass location of the whole body based on differentiation of tissues of a multi-divided trunk. *Journal of Biomechanics*, 48:65-72.

Ferguson, A., Moritz, M. (2016) *Leading. From Life and My Years at Manchester United.* Polish Edition. Krakow: Sine Qua Non.

Hadhazy, A., Weiner, H. R. (2008) What makes Michael Phelps so good? *Scientifican American* August 18, 2018. Available: https://www.scientificamerican.com/article/what-makes-michael-phelps-so-good1/. Accessed March 17, 2019.

Hanavan, E. P. Jr. (1964) *A mathematical model of the human body*. Technical Report AMRL-TR-64-102. Behavioral Sciences Laboratory, Aerospace Medical Research Laboratories, Aerospace Research Division, Air Force System Command, Dayton, Ohio: Wright-Patterson Air Force Base.

Hanlon, M., Smith, J. (2012) London 2012 Olympics: Faster. Higher. Longer. Stronger. *The Telegraph*, August 3, 2012. https://www.telegraph.co.uk/sport/ olympics/9449673/London-2012-Olympics-Faster.-Higher.-Longer.-Stronger.html. Accessed March 18, 2019.

Harless, E. (1857) Static moments of human segments. (in German) *Proceedings of the Mathemathical-Physical Class of the Royal-Bayern Academy of Science*, Munich, 8:71-97.

King, W. F., and Mertz, H. J., eds. (1973) *Human Impact Response. Measurement and Simulation.* Proceedings of the Symposium on Human Impact Response held at the General Motors Research Laboratories, Warren, Mi., October 2-3, 1972, New York: Plenum Press.

Kowalczyk, R. (2013) *Localization of center of mass of high jumpers and its position according to the bar* (in Polish). Doctoral dissertation, Gdansk: Gdansk University of Physical Education and Sport.

Lees, E. (2012) Paralympics 2012: Pistorius vs Oliveira. *The Telegraph*, https://www.telegraph.co.uk/sport/olympics/paralympic-sport/9523026/Paralympics-2012-Pistorius-vs-Oliveira.html. Accessed March 24, 2019.

Levy, L, Mills, G. (2009) Interview with Glenn Mills. *IAAF New Studies in Athletics*, 24:1:29-34.

Riener, R., Kasielke, N. (2016) Technology for Everyday Life. In *Programm/Programme. Cybathlon, 8. Oktober 2016. Swiss Arena Kloten.* Zürich: Eidgenösische Technische Hochschule.

Singla, D. and Veqar Z. (2014) Methods of Postural Assessment Used for Sports Persons. *Journal of Clinical & Diagnostic Research* 8(4):LE01-LE04. Accessed March 24, 2019. doi: 10.7860/JCDR/2014/6836.4266.

Snyder, R. G., Schneider, L. W., Owings, C. L., Reynolds, H. M., Golomb, D. H., Schork, M. A. (1977) *Anthropometry of Infants, Children, and Youth to Age 18 for Product Safety Design.* Final Report UM-HSRI-77-17, Highway Safety Research Institute, Ann Arbor, Mi.: The University of Michigan.

Wood R. (2001) World Best VO2 max scores. In *Topend Sports. The Sports Fitness, Nutrition and Science Resource.* Available: www.topendsports.com. Accessed March 17, 2019.

Zatsiorsky V. M., Seluyanov V. N. (1979) Mass-inertial characteristics of human body segments and their relations with anthropological landmarks (in Russian). *Questions of Anthropology (Voprosy antropologii)*, 62:91-103.

Zatsiorsky V. M., Aruin A. S., Seluyanov V. N. (1981) *Biomechanics of the human movement system* (in Russian). Moscow: Fizkultura and Sport.

BIOGRAPHICAL SKETCH

Wlodzimierz S. Erdmann

Affiliation: Gdansk University of Physical Education and Sport, Gdansk, Poland

Education: 1) Poznan University of Physical Education, Poznan, Poland; 2) Gdansk University of Technology, Gdansk, Poland

Business Address: Gdansk University of Physical Education and Sport, 1 Gorskiego Str., 80-336 Gdansk, Poland

Research and Professional Experience: Biomechanics, Kinesiology, Sports Sciences, Biomedical and Sport Engineering, Forensic Expertises

Professional Appointments: Part time Full Professor at the Gdansk University of Physical Education and Sport, Gdansk, Poland

Honors: Doctor of Philosophy in Physical Culture / Biomechanics, Doctor of Science (Doctor habilitated) in Physical Culture / Biomechanics of Sport

Publications from the Last 3 Years:

Erdmann, W. S. (2016) *Engineering of Movement Rehabilitation. An Outline*. Gdansk: Gdansk University of Technology.

Erdmann W. S. (2016) Biomechanist as an advisor to runners. *Visnik chernigivskogo natsionalnogo pedagogichnogo universitetu*, Chernigiv, Ukraine, 139:1:60-65.

Erdmann W. S., Giovanis V., Aschenbrenner P., Kiriakis V., Suchanowski A. (2017) Methods for acquiring data on terrain geomorphology, course geometry and competitors' runs in alpine skiing: a historical review. *Acta of Bioengineering and Biomechanics*, 19(1):69-79.

Erdmann W. S. (2017) Geometry and inertia of the body as important factors in sport biomechanics. *Visnik chernigivskogo natsionalnogo pedagogichnogo universitetu*, Chernihiv (Ukraina), 147:1:94-100.

Al Sudani A. D., Erdmann W. S. (2017) Development of body morphology of Iraqi students: sedentary way of life and fatness. *Baltic Journal of Health and Physical Activity*, 9:2:30-38.

Erdmann W. S., Dancewicz-Nosko D., Giovanis V. (2018) Velocity distribution of women's 30-km cross-country skiing during Olympic Games from 2002-2014. *The Journal of Sports Medicine and Physical Fitness*. Edizioni Minerva Medica, DOI: 10.23736/S0022-4707.17.07948-8.

Erdmann W. S., Aschenbrenner P., Giovanis V. (2018) Methods for acquiring kinematic data in alpine skiing. Blog, British Journal of Sports Science, Part 1: Sep 23, 2018; Part 2: Oct 15, 2018.

Erdmann W. S. (2018) Equipment and Facilities Adapted for Disabled People in Recreation and Sport. *MOJ Applied Bionics and Biomechanics*, 2:1:11-16.

Erdmann W. S. (2018) Center of mass of the human body helps in analysis of balance and movement. *MOJ Applied Bionics and Biomechanics*, 2:2:144-148.

Erdmann W. S. (2018) Biomechanics and Bionics in Sport. *MOJ Applied Bionics and Biomechanics*, 2:3:200-201.

In: A Closer Look at Biomechanics ISBN: 978-1-53615-866-3
Editor: Daniela Furst © 2019 Nova Science Publishers, Inc.

Chapter 6

CLINICAL APPLICATION OF PEDOBAROGRAPHY

Arno Frigg[*]

Foot and Ankle Surgery Zuerich, Zuerich, Switzerland
Orthopaedic Surgery, University Hospital Basel, Basel, Switzerland

ABSTRACT

Pedobarography is widely used in research, but its application in clinical practice is still at the beginning. There are several reasons for this: (1) The installation of the equipment is time-consuming, which is a strain on the busy clinical schedule. (2) The equipment is relatively expensive, thereby pushing up costs. (3) It is often difficult to draw meaningful clinical conclusions from the large amount of data that pedobarographic exams provide. (4) Laboratory measurements often provide only limited insight into load patterns that feet are exposed to in the daily life of a patient.

These limitations notwithstanding, pedobarography is becoming an increasingly useful and important instrument in the clinical context. In this article we show how the large number of pedobarographic parameters, which varies from 72 to 198 per foot, can be aggregated into

---

[*] Corresponding Author's E-mail: mail@arnofrigg.com.

a single indicative parameter, the so-called Relative Midfoot Index (RMI). This enables that clinicians do not have to analyze hundreds of pedobarographic parameters and instead can reach a meaningful interpretation by focusing only on the RMI, possibly combined with a visual interpretation of force/pressure time graphs in which healthy subjects show a triphasic force-time curve while diseased subjects show a biphasic force-time curve with a flatter midfoot depression. We therefore recommend a standardization of the reporting of pedobarographic outputs in terms of the RMI as well as the Maximal Force (as representation of load) and the contact times (as representation of rollover) in only three areas (hind-, mid- and forefoot).

The article goes on to present clinical results obtained through the use of pedobarophy. (1) We show that contralateral feet cannot be used as a control group for a study of diseased feet because the gait of contralateral feet is significantly different from the gait of healthy feet. (2) Total ankle replacement shows no advantages over ankle arthrodesis when measured in shoes. This is clinically important because we almost always wear shoes, which makes barefoot measurements largely irrelevant. (3) Pedobarographic methods show that hindfoot alignment and the translation of radiological alignment into pedobarographic loading is crucial. This has the important clinical consequence that ankle arthrodesis have to be positioned in an angle of about 10° of valgus and that the standard *in situ* fusions of ankle osteoarthritis, which is typically in varus, yields inferior outcomes.

Finally, we describe the new "third generation" pedobarographic systems and their potential in future research. First generation systems were pressure mats, which are nowadays available in commercial sports and shoe stores. Second generation systems were insole-sensors that measured foot-shoe interface pressures attached to cables in the laboratory. The new third generation systems are insoles with sensors and integrated or external data storage as well as power supply, making it possible to measure the foot-shoe interface over extended periods of time (days and weeks). Data are transmitted via Bluetooth, which makes cables unnecessary.

Keywords: pedobarography, standardized reporting, clinical application, total ankle replacement, ankle arthrodesis

1. INTRODUCTION

Pedobarography is widely used in research, and its colored pictures are so appealing that many sport shoe retailers use the method to sell shoe wear. Nevertheless, the application of pedobarography in clinical practice is still at the beginning. There are several reasons for this. Some reasons concern the practicalities of pedobarography. The installation of the equipment is relatively time-consuming, which is a strain on a busy clinical schedule, and the equipment is still expensive, which pushes treatment costs up. We address this issue in Section 2, where we discuss different kinds of pedobarographic equipment and their application in practice. We identify three different generations of equipment and point out that the practical problems are significantly diminished in the third, most recent, generation.

Other reasons are intrinsic to the method. It is generally difficult to draw meaningful clinical conclusions from pedobarographic data. In part this difficulty is due to the fact that practitioners are unfamiliar both with basic principles of gait and loading and the functioning of the equipment, which can lead to misinterpretation of pedobarographic graphs. For instance, it does not seem to be generally known that many peak pressure values are completely normal (e.g., heel strike in varus, push off over the great toe). In Section 3 we discuss the principles of gait and loading and explain how data from contemporary equipment should be interpreted. However, in part this difficulty is also due to the fact that pedobarographic equipment can output large amounts of data, and it is often difficult to draw meaningful clinical conclusions from these data. Most systems record parameters in 4 - 11 areas with 18 or more parameters per area, which results in 72 - 198 parameters per patient. There is no standardization on which parameters should be reported, and in many publications there seems to be a bias of positive reporting: with this huge amount of parameters the chance is very high that a statistical test will provide a significant result in some parameters with a $p < 0.05$ without correcting for multiple testing.

In Section 4 we suggest that this problem can be overcome by aggregating output data into a single indicative parameter, the so-called Relative Midfoot Index (RMI). Clinicians then do not have to analyze hundreds of pedobarographic parameters and instead can reach a meaningful interpretation by focusing only on the RMI, possibly combined with a visual interpretation of force/pressure time graphs in which healthy subjects show a triphasic force-time curve while diseased subjects show biphasic force-time curve with a flatter midfoot depression (Fig. 1-3). We therefore recommend a standardization of the reporting of pedobarographic outputs in terms of the RMI as well as the Maximal Force (as representation of load) and the contact times (as representation of rollover) in only three areas (hind-, mid- and forefoot).

In Section 5 we describe clinical applications of pedobarography so far in the field of ankle instability, diabetic ulcerations and unloading of the foot in lower leg orthoses. In Section 6 we present clinical results obtained through the use of pedobarophy. (1) We show that contralateral feet cannot be used as a control group for a study of diseased feet because the gait of contralateral feet is significantly different from the gait of healthy feet. (2) Total ankle replacement shows no advantages over ankle arthrodesis when measured in shoes. This is clinically important because we almost always wear shoes, which makes barefoot measurements largely irrelevant. (3) Pedobarographic methods show that hindfoot alignment and the translation of radiological alignment into pedobarographic loading is crucial. This has the important clinical consequence that ankle arthrodesis have to be positioned in an angle of about 10° of valgus and that the standard *in situ* fusions of ankle osteoarthritis, which is typically in varus, yields inferior outcomes.

Finally, Section 7 provides an outlook. It outlines what clinical applications may become available due to pedobarography in the near future with the new "third generation" pedobarographic systems. First generation systems were pressure mats, which are nowadays available in commercial sports and shoe stores. Second generation systems were insole-sensors that measured foot-shoe interface pressures attached to cables in the laboratory. The new third generation systems are insoles with sensors and integrated or external data storage as well as power supply, making it possible to measure the foot-shoe interface over extended periods of time (days and weeks). Data are transmitted via Bluetooth, which makes cables unnecessary.

2. PEDOBAROGRAPHIC EQUIPMENT

Pedobarographic equipment has been developed for number of years and different products have been brought on the market. To aid orientation we propose to classify these products as belonging to three different generations.

First generation pedobarography products are flat mats for a patient to walk on. Today they are widely available for a price of a few thousand dollars (e.g., www.novel.de, www.tekscan.com, www.megascan.de). Because of the acceleration and deceleration at the beginning and the end of the sequence of steps, subjects should make 3-5 steps before and after stepping onto the pressure mat. For this reason such a runway has to be around 6-10 meters long. This is more space than a usual doctor's office or hospital consultation room can offer, and hence the runway requires a separate laboratory room. Carrying out a test is time consuming because the patient needs to be instructed and at least 3 - 5 measurements for each foot have to be made in order to average measurements. All in all this will take between 20 and 30 minutes per patient. Despite these efforts, the results are of limited use. Measurements on the matt are done barefoot, but since people almost always wear shoes the measurement results provide only limited insight into load patterns that feet are exposed to in the daily life of a patient. To get meaningful data, the unit consisting of both the foot and the shoe together should be biomechanically investigated and assessed.

Second generation pedobarographic systems are insoles that are placed in the shoe and measure the interface between foot and shoe (e.g., www.novel.de, www.tekscan.com, www.megascan.de). These systems deliver much more realistic data than the first generation mats, concerning the forces that act on the foot during walking, running or stop-and-go sports. However these systems are expensive with prices ranging from tens of thousands to hundreds of thousands of dollars. They have to be installed in a laboratory setting by a qualified research assistant. They are connected to the data recorder by cable, and depending on the model that amount of cables required can be considerable. The battery capacity is generally limited to 10-30 minutes, with some specific systems reaching a maximum of up to two hours. For these reasons, these systems can only be used in a laboratory setting, which provides only limited information about how patients would walk in their everyday lives. Testing one patient takes about 1-2 hours, which is considerable.

Recently third generation of measuring systems have become available. With costs ranging between 300 and 1500 dollars per unit, these systems are considerably cheaper than first and second generation systems (e.g., www.golex.ch, www.moticon.de). These new systems are basically insoles with sensors and integrated or external data storage as well as power supply, making it possible to measure the foot-shoe interface over extended periods of time (days and weeks). Data are transmitted via Bluetooth, which makes cables unnecessary. Third generation equipment is at once technically superior and more affordable than first and second generation equipment. However, one of the fundamental problems of pedobarography remains unresolved: what meaningful clinical decisions can be based on these data. We turn to this problem in Section 4 where we will outline a new method of aggregating data into one indicative index.

3. PRINCIPLES OF GAIT

In this section we briefly review the principles of gait, which are essential to a correct interpretation of pedobarographic results.

There is the stance phase and the swing phase. In pedobarography, only the forces that result from the interaction of the foot with the ground during stance phase are measured. It is normal that the heel contacts the ground first on the lateral side in varus, and then the hindfoot is unlocked by moving the heel into slight valgus for the rollover process. Then the push off takes place, normally over the great toe and first metatarsal head. For this reason it is normal to have peak pressures in these areas.

The loading of the medial half of the foot in patients with healthy feet is higher than the loading of lateral half of the foot [21, 22]. So it is normal to have an increased medial load and this is not a pathologic sign for flat foot. Findings on healthy subjects show 33% more load medially related to the anatomic axis of the foot [21, 22]. Other authors assumed a 50:50 medial-to-lateral force distribution for intraoperative pedobarography [47], but this is not correct.

These points are important to bear in mind when interpreting visual representations of pedobarographic measurements. Typical measurements-systems do not report pressures of more than 300 kPa and hence and values above that threshold are not displayed. Such values can arise on the side of an ulceration, but also in healthy patients with no problems. For this reason, the dark violet areas on a pedobarographic report (indicating maximal pressure) do not help to identify areas at risk more than a clinical exam of callused skin or the already existence of an ulcer.

Walking speed, usually not at all measured by pedobarography, is also a crucial parameter: The walking speed for a healthy person is about 1.2 m/sec [23]. With ankle osteoarthritis, ankle arthrodesis (AA) or total ankle replacement (TAR), the walking speed decreases to around 0.8 m/sec, and it further decreases to 0.6 m/sec with tibio-talo-calcaneal arthrodesis (TTC) [23]. Research has even found that a slower walking speed is a predictor of mortality: The correlation of overall mortality with walking speed has been analysed in older people, people with peripheral vascular disease, and people with cardio-vascular disease. Meta-analyses show that walking speeds of more than 0.7 m/sec (better even > 1 m/sec or more) show a decrease in the risk of mortality by about 90% compared to slower walking speed of < 0.3 m/sec (or also below 0.7 m/sec) [31, 57]. About a 10% increase of risk of mortality was detected per 0.1m/sec less walking speed [31, 57].

4. DATA-REDUCTION AND STANDARDIZATION OF PEDOBAROGRAPHIC REPORTING

4.1. Standardized Reporting

Pedobarography is widely used to investigate several different pathologies and outcomes after surgery. However, the analysis of the measured data is a challenge: a standard pedobarography system such as the Novel Emed m/E system measures 18 basic and a number of optional parameters in 3 to 10 areas of interest as well as for the total foot. Such a measurement provides values of 72-198 (4 to 11 times 18) parameters for each foot. These raw data are difficult to interpret, and left unprocessed they provide no useful information. Several authors have therefore focused attention on selected parameters such as average pressure, peak pressure, pressure time integral and contact time [5, 8, 44, 51]. However it remains unclear whether selectively focusing on one of these parameters is appropriate because no clear arguments have been given as to how parameters are chosen and why some are privileged over others. This is a serious concern because with so much choice there is a temptation to cherry pick parameters to yield the desired result.

A solution to this problem has to meet several conditions. First, a clear prescription has to be given about which parameters are to be reported. Second the number of parameters has to be reduced because statistically comparing groups of parameters to each other with the full set of 72-198 parameters would require an unmanageable number of tests: for two groups 72-198 tests, for 3 groups 216-594 tests, for four groups 432-1188 tests and so on. In practice this would not only be extremely time consuming; it would also lead to the reporting of many false positive results (if a 5% threshold is assumed, then one would have to reckon with 22-59 false significant test results when comparing 4 groups). Third, load parameters need to be normalized to body weight as any load parameter is directly dependent on the body weight, which distorts comparison between individuals. Forth, also the walking speed influences the ground reaction forces as any force is defined as weight multiplied by acceleration/deceleration (push off/landing phase).

To make pedobarography more user-friendly in a clinical context, we introduced a new parameter called the Relative Midfoot Index (RMI) [19, 20]. The large number of pedobarographic parameters was reduced to 27 – 9 each for hindfoot, midfoot and forefoot – which were then aggregated into two clusters: a cluster of "rollover parameters" describing the temporal motion of the foot over the ground from heel strike to toe off (containing the center of pressure velocity, contact time, instants of maximal force, and instant of peak pressure) and a cluster of "load parameters" (Maximal Force (MF), Peak Pressure (PP), the integral of MF, and the integral of PP). This reduction was essential in order to make the data amenable to statistical analysis and to pave the ground for a clinical interpretation of results. The core results were that the cluster of load for the midfoot was the most important predictor to distinguish between healthy volunteers, patients with AA and TTC, and that the MF had the strongest correlation within this cluster.

The MF is the parameter that provides the most insight into gait mechanics because, unlike pressure, it is independent of local foot callosities or deformities. Therefore the new parameter was based on the MF representing the depth of the midfoot valley of the force-time curve in relation to the amount of MF in the hindfoot and forefoot:

$$RMI = 1 - \frac{2 MF_m}{MF_f + MF_h},$$

where MF_m, MF_f, and MF_h are the MF for the midfoot, forefoot, and hindfoot respectively. In effect the RMI is the MF_m weighted by the average of MF_f and MF_h: in normal triphasic gait the RMI is expected to be close to one (deep midfoot depression on force-time graphs) while in pathologic biphasic gait it is expected to be close to zero (flat midfoot depression on force-time graphs, Figure 1).

The RMI has the advantage that it is independent of body weight and walking speed, which both influence the absolute values. The RMI therefore allows for simple comparisons between individuals (Figure 2). The RMI is larger for healthy feet and smaller for diseased feet. As regards cut-off values, results of ankle osteoarthritis (RMI = 0.65) and healthy people (RMI = 0.84) showed that an RMI of 0.8 was the best compromise to determine whether a foot is healthy or not, with a positive predictive value of 80% and sensitivity of 78%. This observation was also made after calcaneal fractures where a flattened ground reaction force was found [37]. We suspect that the flattened ground reaction force in the midfoot is an indicator for a pathologic gait independent of the underlying disease. However, further research will be needed to confirm this.

In sum, the large number of pedobarographic parameters can be reduced to the interpretation of the relative midfoot index and the visual interpretation of force/pressure time graphs. In addition, for pedobarographic research we recommend reporting the MF as representation of load and the contact times as representation of rollover in three areas (hindfoot, midfoot and forefoot).

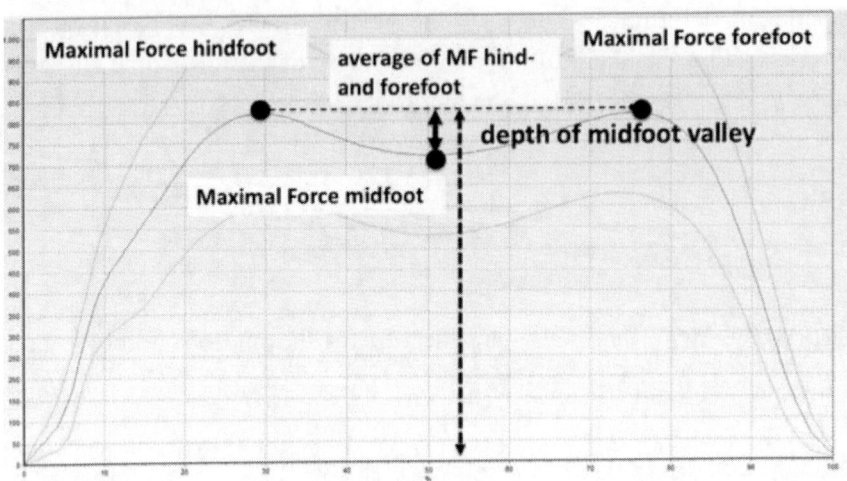

Figure 1. Relative Midfoot Index (RMI): The RMI is calculated by putting the depth of the midfoot valley in relation to the average of the Maximal Force in the hind- and forefoot.

Clinical Application of Pedobarography

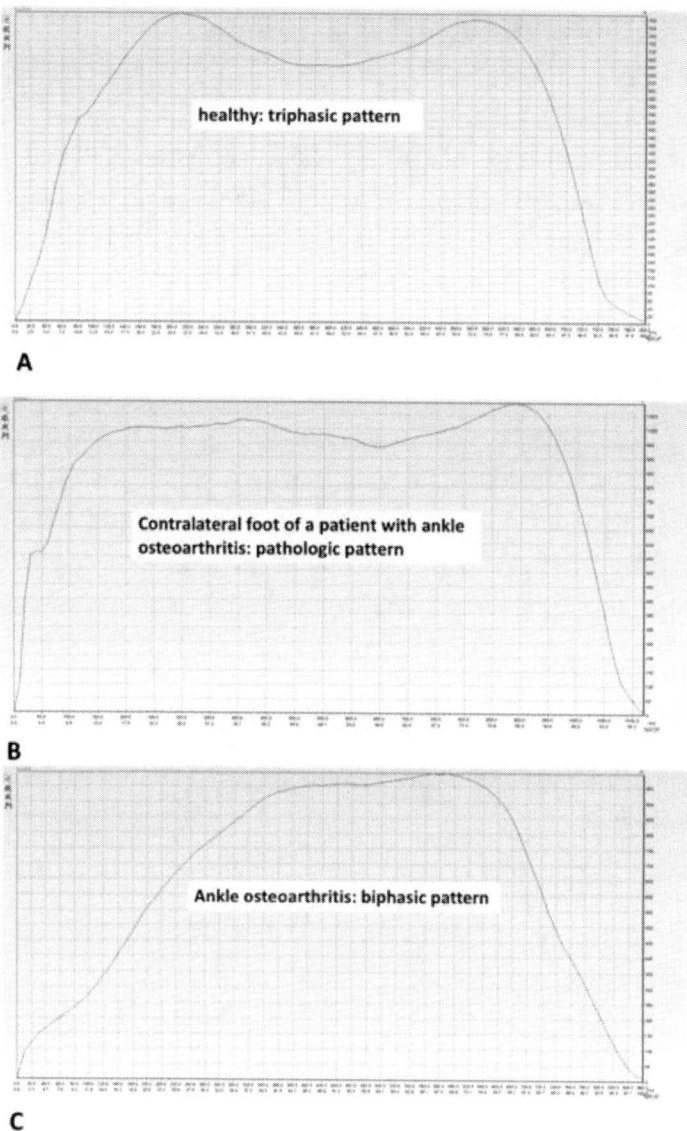

Figure 2. Maximal force curves of a healthy foot (A), of the contralateral unaffected foot of a patient with ankle osteoarthritis (B) and a foot with ankle osteoarthritis (C). These graphs show that the midfoot depression is small in the ankle-osteoarthritic foot (biphasic pattern) and deep in the healthy foot (triphasic pattern). Figure b shows that the unaffected foot has a pathologic gait pattern and is therefore not suitable for comparison.

4.2. Comparison of Diseased Feet to Healthy and Contralateral Feet

Another problem one encounters when interpreting pedobarographic data is that there are two different ways of comparing affected feet: either comparing a patient's affected foot with the patient's unaffected contralateral foot, or the comparing the affected foot with feet of other healthy participants. The first method assumes that contralateral feet can be regarded as healthy and that the foot problem of one side would not affect the other side. This assumption has been proven to be wrong (Figure 2) [19, 20, 23]. Contralateral feet of patients with an affected foot exhibited an RMI of 0.7 and a pathologic force-time curve. In fact, contralateral feet turned out to be highly significantly different ($p < 0.0001$) from healthy feet in almost all parameters, the exception being the MF hindfoot. This clearly shows that contralateral feet cannot be taken as comparisons [19, 20, 23].

4.3. Association of Pedobarographic Data to Clinical Outcome Scores

We found a positive correlation between RMI and AOFAS-score ($r = 0.48$) [19, 20, 23]. In the past, other studies have also tried to identify correlations between AOFAS score and pedobarographic parameters. Schuh found a significant correlation between the AOFAS score and the loading parameters of the medial midfoot after surgery of posterior tibial tendon dysfunction [52]. However the reported correlation coefficients were between -0.29 and -0.36, which are values that we consider too weak. Rammelt realized, that a higher AOFAS-score is associated with a higher pressure time integral in the whole foot, but without reporting a correlation coefficient [44]. Burns noticed a correlation between the pressure time integral and foot pain ($r = 0.49$) in cavovarus feet [8]. In rheumatoid feet, Schmiegel detected an increase of the average pressure together with the severity of the impairment in the Health Assessment Questionnaire in three groups, but no correlation coefficient was calculated [51]. In sum, the literature only reports weak correlations between pedobarographic parameters and clinical findings.

5. CLINICAL APPLICATIONS OF PEDOBAROGRAPHY

The largest problem for pedobarography is that its systems are usually used in laboratory settings and that not enough efforts have been made to apply the systems in a clinical context. For clinical applications, pedobarographic systems were combined with a biofeedback detector (e.g., PEDAR-system of Novel). For example an acoustic or visual signal was measured in case a loading over the adjusted threshold in a certain area. Such biofeedback systems were used in to investigate proprioception, diabetic ulcerations, and the unloading of the foot and ankle after surgery or injury.

5.1. Proprioception

Biofeedback-systems with a pedobarographic measurement of the pressure distribution on the foot were used to improve upright stand with biofeedback on the back in a laboratory setting with 10-90 sec tests [58, 63]. Another application was a warning system to avoid ankle sprains. In a laboratory setting the PEDAR-system with 30 sec trials was used to give an acoustic alarm when too much load was put on the lateral foot [16].

5.2. Diabetic Ulcerations

In case of diabetic neuropathy, the removal of peak-pressure is crucial to heal diabetic ulcerations. Several researchers have therefore developed different kinds of biofeedback systems with either acoustic alarm, visual biofeedback, or analysis for measurement together with the patient on the screen to reduce pressure on the ulcerations area [13, 14, 17, 18, 42, 50]. In most studies, the PEDAR-system of Novel was used. Some researchers have measured the load for 15min and the patients received the feedback in order to change the kind of gait [50]. Others have trained the patients on day one, then again after 5 and 10 days, to walk differently with 10-15 steps to alleviate the peak pressures on the ulcerations [13, 42].

5.3. Unloading of the Foot and Ankle after Surgery or Injury

After an injury to the foot and ankle or after surgery, usually a partial weight bearing is needed to aid healing of the injured structures. The effect on gait wearing lower leg orthosis was investigated [41]. Patients were observed on a 10m runway and it was found that the load on the heel was increasing as patients had to unload more weight.

6. SCIENTIFIC ADVANCEMENTS BY PEDOBAROGRAPHY

6.1. Ankle Arthrodesis versus Total Ankle Replacement

Chronic instability or ankle fractures can eventually turn into ankle osteoarthritis because of cartilage damage in ankle sprains. The most common treatment options are ankle arthrodesis (AA) or total ankle replacement (TAR). There is no agreement about which method is superior, and the relative merits of the methods are the subject matter of scientific controversy [2, 4, 10, 12, 15, 25-27, 30, 36, 39, 43, 53, 56, 62]. At first sight one would expect the mobile TAR to be superior to the stiff AA. However, a review of the scientific literature comparing TAR and AA reveals a rather more complex picture. (1) AA and TAR have similar postoperative clinical outcomes, which in both cases are superior to patients' preoperative conditions as indicated by improved pain scores and functional scores (AOFAS) [2, 4, 10, 12, 15, 25-27, 30, 3 6, 39, 43, 53, 56, 62]. (2) Both treatments result in the same walking speed, which, however, is slower than in healthy subjects [4, 15, 56]. (3) There is a development of subtalar osteoarthritis (3% in 5 years for AA, 1% in 5 years for TAR) [53]. (4) There is an increased motion of the knee joint to compensate for the rigid ankle. This can lead to a consequent development of arthritis both in AA and TAR, which, however, is discussed controversially [15, 25, 43]. The only obvious advantage of TAR over AA measured with gait analysis was a more symmetrical gait [15, 39].

A different picture emerges when we consider longevity. The revision rate in AA is 7 – 26% compared to 17 – 54% in TAR [12, 30, 53]. Additionally, there is an implant failure in TAR of 24 – 11% after 10 years while AA last forever, which must be taken into account in an assessment [26, 27, 36, 54, 62]. There are only few studies of the treatment effects of tibiotalocalcaneal arthrodesis (TTC) [1, 28, 55]. They report satisfaction scores of 91% for AA and 88% for TTC and good clinical and functional results for both AA and TTC [1, 28, 55]. However, these figures conceal the clinically observed impairment after adding a subtalar fusion to an AA.

6.2. Comparison with Shoes

It is customary to perform gait analysis with subjects going barefoot. This limits the practical relevance of the results because in their everyday lives people usually walk in shoes. We carried out the first ever pedobarographic evaluation of patients after TAR, AA and TTC while they were wearing different shoes [23]. We compared 126 patients (28 TAR, 57 AA, 41 TTC) with 35 healthy volunteers in three situations (barefoot, with running shoes and with rocker bottom shoes) and found no differences between AA or TAR in any of the parameters characterizing the three different circumstances. Patients with AA or TAR were found to walk significantly slower than healthy controls, and they showed an increased maximal force in the forefoot. This is important because that force could be a trigger for adjacent osteoarthritis. We found that running shoes were beneficial for all patients, but rocker bottom shoes in fact provided little added benefit. From this we conclude that AA is equal to TAR, but both are inferior to healthy controls. It therefore remains unclear whether TAR really enjoys a benefit over AA, in particular given the high rates of failure and revision of TAR [23].

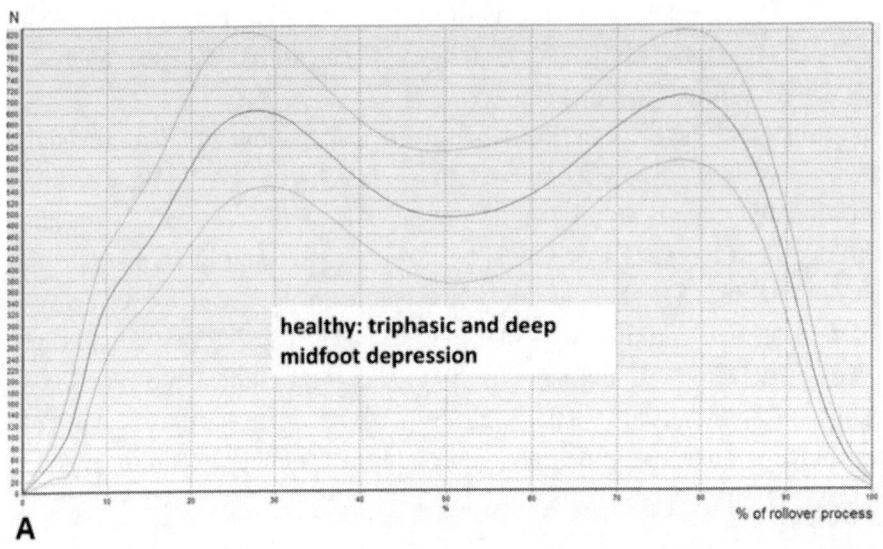

A

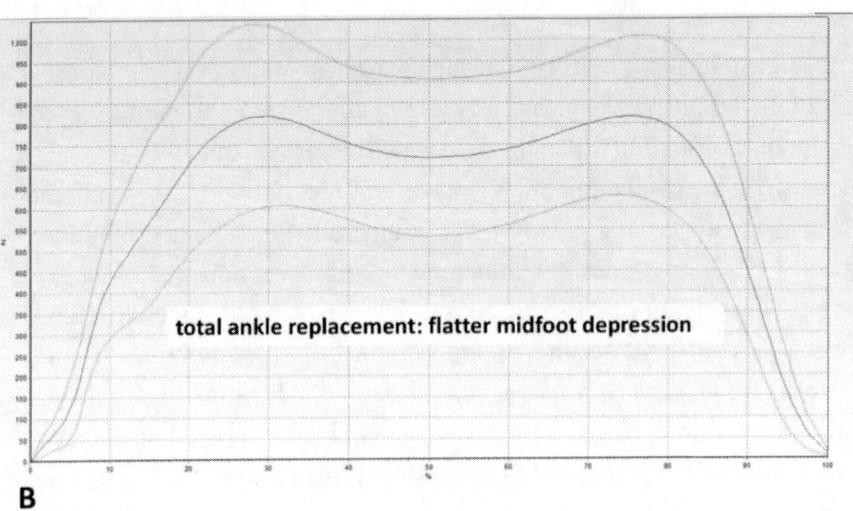

B

Figure 3. (Continued).

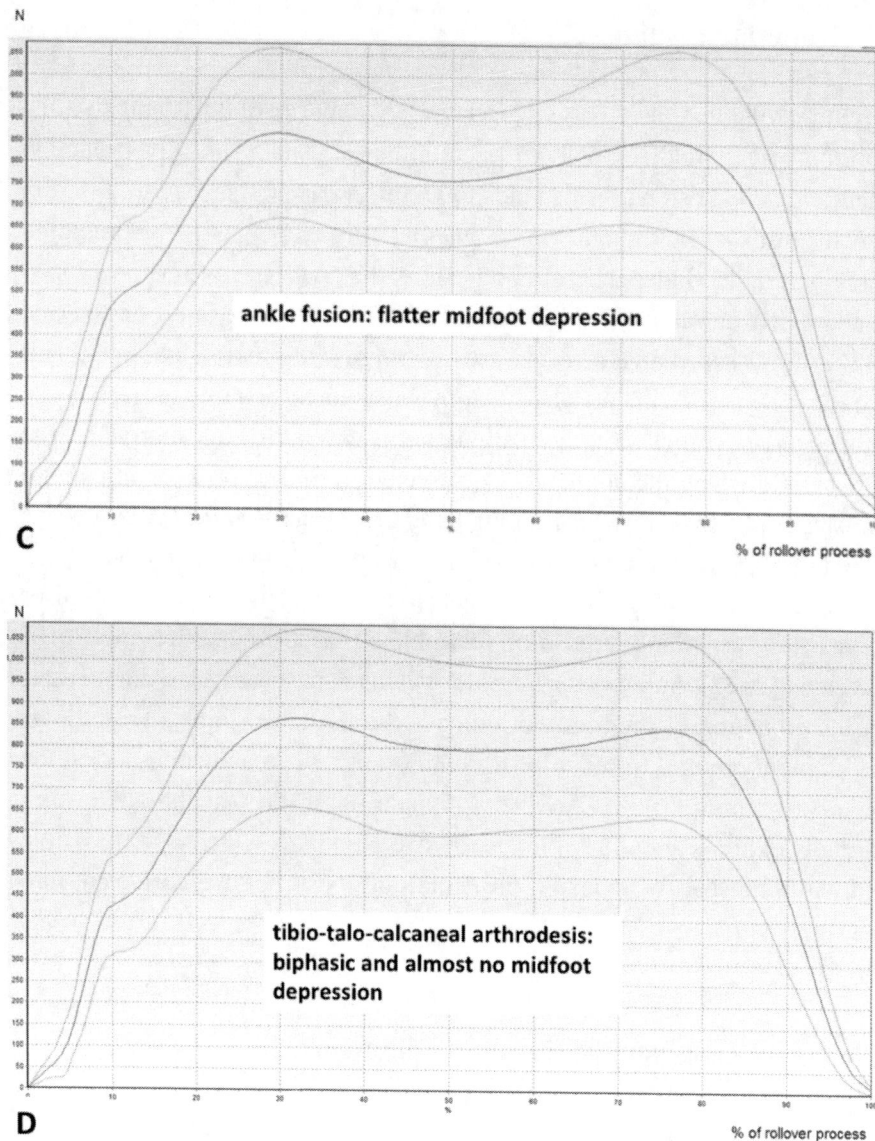

Figure 3. Maximal force curves a healthy volunteers (A), patient with total ankle replacement (B), ankle arthrodesis (C) and tibial-talo-calcaneal arthrodesis (D). The more extensive the fusion, the less deep the midfoot depression. (These graphs were directly produced by the Novel m/E software and show the mean and 95% confidence interval of 35 healthy volunteers, 28 total ankle replacements, 57 patients with ankle and 42 patients with TTC-arthrodeses).

6.3. Hindfoot Position

In the ongoing discussion over the relative merits of TAR and AA, the position of the hindfoot of both operations are usually not given enough attention. Ankle osteoarthritis tends to go into varus. Often AA are fused *in situ* in varus even though we know that this can lead to an impaired outcome [7, 22]. Independent reports show that AAs have to be positioned in about 10° of valgus [7, 22]. The more the hindfoot position of AA and TTC-arthrodeses changed from varus to valgus, the more similar the gait pattern measured with pedobarography became to healthy subjects [22]. This contrasts with TAR, where alignment is crucial for the survival of the implant and where the alignment is corrected to neutral position in either a one-stage or a two-stage operation. To evaluate the hindfoot position, we recommend either the hindfoot-alignment-view or the long-axial view [22]. The hindfoot position of healthy subjects is in 1 - 6° of varus, which raises the question why the best position for ankle arthrodesis should be in 0 - 10° of valgus [7, 22]. Support for hindfoot valgus is based on view that valgus causes pronation of the foot that unlocks the midtarsal joints. This allows for a more natural motion of the hindfoot, whereas varus locks the midtarsal joints [33, 38]. Another rationale might be that the ankle joint normally exhibits about 5° of inversion/eversion [32], which is not possible with arthrodesis. As a result the hindfoot must be overcorrected into valgus.

6.4. Visual Judgment of Hindfoot Alignment

The analysis builds on translating radiographic static alignment into dynamic pedobarographic loading [21, 22]. There are correlations of radiographs to pedobarographic measurement with a correlation coefficient ranging approximately from 0.4 to 0.6 [21]. Visual judgment only predicted radiographic alignment in 48% [21]. The inaccuracy of visual judgment has also been reported by other authors [3, 48, 49]. Although the reliability of clinical measurements is improved through the use of a goniometer and a weight-bearing position, clinical approaches are less reliable than radiographic approaches. There was no correlation between visual alignment and any of the pedobarographic parameters [21, 22]. In earlier studies we identified the position of the heel in the hindfoot alignment view according to Saltzman as a crucial factor for the determination of the functional outcome of total ankle replacement, ankle arthrodesis and tibiotalocalcaneal arthrodesis [29, 35]. Based on these findings, a *Hindfoot Alignment Guide* was developed [24], which measures the intraoperative position of the heel. This guide significantly improved the intraoperative positioning compared to visual techniques with an accuracy of 5°. It was especially useful in the context of multilevel corrections in which the need for the amount of a simultaneous osteotomy had to be evaluated intraoperatively. The number of osteotomies could thereby be tailored to the specific needs of the case, either reducing or adding osteotomies [24]. We also evaluated the hindfoot position with an MRI and found only a moderate correlation with the Saltzman view and a tendency to valgisation of the results [6].

7. OUTLOOK

New third generation pedobarographic systems make measurement of the foot-shoe interface in daily life possible (e.g., www.golex.ch, www.moticon.de). This constitutes considerable progress, but the usefulness of these systems is still limited by the capacity of the batteries. Early systems had sufficient energy for one day, after which the battery had to be reloaded. Since most patients who need biofeedback are elderly people who tend not to be used to maintaining technical equipment and recharging batteries (or indeed have complained or refused to do so in our experience), battery power was a serious limitation in practice. More recent models therefore have a battery that lasts for around 7-10 days. This allows the system to guarantee uninterrupted biofeedback to the patient over an extended period of time.

The measuring sensor is placed under the foot in the shoe or orthosis and then connected with a small cable to the computer and battery unit. Modern units have approximately the size like a cycling computer or the computer and battery are directly integrated in an insole. Such systems could improve patient outcomes in the future and also record patients' compliance after surgery.

Almost all injuries to ligaments, tendons, bone or cartilage around the foot and ankle require an unloading period of 6-12 weeks with partial weight-bearing. If the patient is loading too much, the injured structures cannot heal properly. One of the main problems after foot and ankle surgeries are nonunions of the bones. These nonunions are unfortunately very frequent with 5 - 40% depending on the involved joints: ankle joint 18 - 35%, subtalar joint 12 - 24%, talo-navicular joint 9 - 37%, calcaneo-cuboidal joint 8%, Lisfranc joint 9% [9, 11, 34, 40, 45, 46, 60, 61]. Such nonunions can necessitate an extension of the treatment time or unloading time, and as a result can prolong the time off work by several months and may lead to complex and expensive revision surgery. These nonunions occur despite standardized operative techniques, new implants, and an increased optimization of patient-related influencing factors (smoking, nutrition, age, osteoporosis). To improve this undesirably high nonunion rate, the industry pushes the use of expensive growth factors (e.g., DBX, PDGF from 500 - 2000 $) in every surgery to prevent nonunions. Others studies however show that in case of any injury to the bones the body produces an increased quantity of such growth factors all by itself, and that the effect of these additional growth factors is scientifically questionable. A further problem is that many health systems simply do not have the financial means to use these expensive growth factors in every patient.

Looking at the problem of nonunions from a biomechanical point of view reveals that these growth factors indeed do not play a crucial role as long as one keeps in mind that the loading on the injured bones in case of non-compliance can be as much as 15 - 20 times body weight (1.5 - 2 tons) instead of the partial weight bearing as prescribed by the physician. With the new third generation pedobarographic biofeedback systems, the unloading of the foot and ankle could be monitored, and an improvement of the nonunion rate can be expected in the future.

REFERENCES

[1] Ajis, A; Myerson, Tan KJMS. Ankle arthrodesis vs TTC arthrodesis: patient outcomes satisfaction, and return to activity. *Foot Ankle Int.*, 2013, 34 (5), 657–665 (May).

[2] Atkinson, HD; Daniels, TR; Klejman, S; Pinsker, E; Houck, JR; Singer, S. Pre- and postoperative gait analysis following conversion of tibiotalocalcaneal fusion to total ankle arthroplasty. *Foot Ankle Int.*, 2010, 31 (10), 927–932 (Oct).

[3] Backer, M; Kofoed, H. Passive ankle mobility: clinical measurement compared with radiography. *J Bone Joint Surg Br.*, 1989, 71, 696–698.

[4] Beyaert, C; Sirveaux, F; Paysant, J; Molé, D; André, JM. The effect of tibio-talar arthrodesis on foot kinematics and ground reaction force progression during walking. *Gait Posture*, 2004, 20 (1), 84–91 (Aug).

[5] Bosch, K; Nagel, A; Weigend, L; Rosenbaum, D. From "first" to "last" steps in life – pressure patterns of three generations. *Clin Biomech.*, 2009, 24(8), 676-81.

[6] Büber, N; Zanetti, M; Frigg, A; Mamisch, N. Evaluation of the apparent moment arm on non weight bearing coronal MRI images. *Accepted 7/2017 in Skeletal Radiology.*, 2018, 47(1), 19-24.

[7] Buck, P; Morrey, BF; Chao, EY. The optimum position of arthrodesis of the ankle: a gait study of the knee and ankle. *J Bone Joint Surg Am.*, 1987, 69, 1052–1062.

[8] Burns, J; Crosbie, J; Hunt, A; Ouvrier, R. The effect of pes cavus on foot pain and plantar pressure. *Clin. Biomech.*, 20(9), 877-882, 2005.

[9] Carranza-Bencano, A; Tejero, S; Fernández Torres, JJ; Del Castillo-Blanco, G; Alegrete-Parra, A. Isolated talonavicular joint arthrodesis through minimal incision surgery. *Foot Ankle Surg.*, 2015 Sep, 21(3), 171-7.

[10] Coester, LM; Saltzman, CL; Leupold, J; Pontarelli, W. Long-term results following ankle arthrodesis for post-traumatic arthritis. *J. Bone Joint Surg. Am.*, 2001, 83-A (2), 219–228 (Feb).

[11] Coughlin, MJ; Mann, RA; Saltzman, CL. Surgery of the foot and ankle. Mosby, Inc. Elsevier. Philadelphia, 2008.

[12] Daniels, TR; Younger, AS; Penner, M; Wing, K; Dryden, PJ; Wong, H; Glazebrook, M. Intermediate-term results of total ankle replacement and ankle arthrodesis: a COFAS multicenter study. *J. Bone Joint Surg. Am.*, 2014, 96 (2), 135–142 (Jan 15).

[13] De León Rodriguez, D; Allet, L; Golay, A; Philippe, J; Assal, JP; Hauert, CA; Pataky, Z. Biofeedback can reduce foot pressure to a safe level and without causing new at-risk zones in patients with diabetes and peripheral neuropathy. *Diabetes Metab Res Rev.*, 2013 Feb, 29(2), 139-44.

[14] Descatoire, A; Thévenon, A; Moretto, P. Baropodometric information return device for foot unloading. *Med Eng Phys.*, 2009 Jun, 31(5), 607-13.

[15] Doets, HC; van Middelkoop, M; Houdijk, H; Nelissen, RG; Veeger, HE. Gait analysis after successful mobile bearing total ankle replacement. Foot Ankle Int., 2007, 28 (5), 313–322 (May).

[16] Donovan, L; Feger, MA; Hart, JM; Saliba, S; Park, J; Hertel, J. Effects of an auditory biofeedback device on plantar pressure in patients with chronic ankle instability. *Gait Posture.*, 2016 Feb, 44, 29-36.

[17] Femery, VG; Moretto, PG; Hespel, JM; Thévenon, A; Lensel, G. A real-time plantar pressure feedback device for foot unloading. *Arch Phys Med Rehabil.*, 2004 Oct, 85(10), 1724-8.

[18] Femery, V; Potdevin, F; Thevenon, A; Moretto, P. Development and test of a new plantar pressure control device for foot unloading. Ann Readapt Med Phys., 2008 May, 51(4), 231-7.

[19] Frigg, A; Schäfer, J; Dougall, H; Rosenthal, R; Valderrabano, V. The midfoot load shows impaired function after ankle arthrodesis. *Clin. Biomech.*, 27(10), 1064-1071, 2012.

[20] Frigg, A; Frigg, R; Wiewiorski, M; Goldoni, J; Horisberger, M. Facilitating the interpretation of pedobarography: the relative midfoot index as marker for pathologic gait in ankle osteoarthritic and contralateral feet. *J Foot Ankle Res.*, 2016, Dec 1, 9, 47. §

[21] Frigg, A; Nigg, B; Hinz, L; Valderrabano, V; Russell, I. Clinical relevance of hindfoot alignment view in total ankle replacement. *Foot Ankle Int.*, 2010 Oct, 31(10), 871-9.

[22] Frigg, A; Nigg, B; Davis, E; Pederson, B; Valderrabano, V. Does alignment in the hindfoot radiograph influence dynamic foot-floor pressures in ankle and tibiotalocalcaneal fusion? *Clin Orthop Relat Res.*, 2010 Dec, 468(12), 3362-70.

[23] Frigg, A; Frigg, R. The influence of footwear on functional outcome after total ankle replacement, ankle arthrodesis, and tibiotalocalcaneal arthrodesis. *Clin Biomech* (Bristol, Avon)., 2016 Feb, 32, 34-9.

[24] Frigg, A; Jud, L; Valderrabano, V. Intraoperative positioning of the hindfoot with the hindfoot alignment guide: a pilot study. *Foot Ankle Int.*, 2014 Jan, 35(1), 56-62.

[25] Hahn, ME; Wright, ES; Segal, AD; Orendurff, MS; Ledoux, WR; Sangeorzan, BJ. Comparative gait analysis of ankle arthrodesis and arthroplasty: initial findings of a prospective study. *Foot Ankle Int.*, 2012, 33 (4), 282–289 (Apr).

[26] Henricson, A; Skoog, A; Carlsson, A. The Swedish Ankle Arthroplasty Register: an analysis of 531 arthroplasties between 1993 and 2005. *Acta Orthop.*, 2007, 78 (5), 569–574 (Oct).

[27] Hobson, SA; Karantana, A; Dhar, S. Total ankle replacement in patientswith significant pre-operative deformity of the hindfoot. *J. Bone Joint Surg.* (Br.), 2009, 91 (4), 481–486 (Apr).

[28] Jastifer, J; Coughlin, MJ; Hirose, C. Performance of total ankle arthroplasty and ankle arthrodesis on uneven surfaces, stairs, and inclines: a prospective study. *Foot Ankle Int.*, 2015, 36 (1), 11–17 (Jan).

[29] Johnson, JE; Lamdan, R; Granberry, WF; et al. Hindfoot coronal alignment: a modified radiographic method. *Foot Ankle Int.*, 1999, 20(12), 818-825.

[30] Krause, FG; Windolf, M; Bora, B; Penner, MJ; Wing, KJ; Younger, AS. Impact of complications in total ankle replacement and ankle arthrodesis analyzed with a validated outcome measurement. *J. Bone Joint Surg. Am.*, 2011, 93 (9), 830–839, (May 4).

[31] Liu, B; Hu, X; Zhang, Q; Fan, Y; Li, J; Zou, R; Zhang, M; Wang, X; Wang, J. Usual walking speed and all-cause mortality risk in older people: A systematic review and meta-analysis. *Gait Posture.*, 2016 Feb, 44, 172-7.

[32] MacWilliams, BA; Cowley, M; Nicholson, DE. Foot kinematics and kinetics during adolescent gait. *Gait Posture.*, 2003, 17, 214-224.

[33] Mann, RA. Surgical implications of biomechanics of the foot and ankle. *Clin Orthop Relat Res.*, 1980, (146), 111-118.

[34] Miehlke, W; Gschwend, N; Rippstein, P; Simmen, BR. Compression arthrodesis of the rheumatoid ankle and hindfoot. *Clin Orthop Relat Res.*, 1997 Jul, (340), 75-86.

[35] Miller, M. *Review of Orthopaedics*. Philadelphia, PA: Saunders, Elsevier, 2004.

[36] Mittlmeier, T. Arthrodesis versus total joint replacement of the ankle. *Unfallchirurg*, 2013, 116 (6), 537–550 (Jun).

[37] Mitternacht, J; Lampe, R. Ermittlungfunktionellerkinetischer Parameter aus der plantaren Druckverteilungsmessung [Determination of functional kinetic parameters from the plantar pressure distribution measurement]. *Z Orthop.*, 144(4), 410-418, 2006.

[38] Morris, JM. Biomechanics of the foot and ankle. *Clin Orthop Relat Res.*, 1977, 122, 10–17.

[39] Müller, S; Wolf, S; Döderlein, L. Three-dimensional analysis of the foot following implantation of a HINTEGRA ankle prosthesis: evaluation with the Heidelberg foot model. *Orthopade*, 2006, 35 (5), 506–512 (May).

[40] Neuman, J; Weinberg, M; Saltzman, C; Barg, A. Does open versus arthrosopic surgical technique affect the union rate of tibiotalar arthrodesis? *AOFAS-Conference*, Boston, 2018.

[41] North, K; Potter, MQ; Kubiak, EN; Bamberg, SJ; Hitchcock, RW. The effect of partial weight bearing in a walking boot on plantar pressure distribution and center of pressure. *Gait Posture.*, 2012 Jul, 36(3), 646-9.

[42] Pataky, Z; de León Rodriguez, D; Allet, L; Golay, A; Assal, M; Assal, JP; Hauert, CA. Biofeedback for foot offloading in diabetic patients with peripheral neuropathy. *Diabet Med.*, 2010 Jan, 27(1), 61-4.

[43] Piriou, P; Culpan, P; Mullins, M; Cardon, JN; Pozzi, D; Judet, T. Ankle replacement versus arthrodesis: a comparative gait analysis study. *Foot Ankle Int.*, 2008, 29 (1), 3–9 (Jan).

[44] Rammelt, S; Grass, R; Zawadski, T; Biewener, A; Zwipp, H. Foot function after subtalar distraction bone-block arthrodesis: a prospective study. *J. Bone Joint Surg. Br.*, 86-B(5), 659-668, 2004.

[45] Rammelt, S; Marti, RK; Zwipp, H. Arthrodesis of the talonavicular joint. *Orthopade.*, 2006 Apr, 35(4), 428-34.

[46] Rammelt, S; Schneiders, W; Schikore, H; Holch, M; Heineck, J; Zwipp, H. Primary open reduction and fixation compared with delayed corrective arthrodesis in the treatment of tarsometatarsal (Lisfranc) fracture dislocation. *J Bone Joint Surg Br.*, 2008 Nov, 90(11), 1499-506.

[47] Richter, M; Frink, M; Zech, S; et al. Intraoperative pedography: a validated method for static intraoperative biomechanical assessment. *Foot Ankle Int.*, 2006, 27(10), 833-842.

[48] Saltzman, CL; Nawoczenski, DA; Talbot, KD. Measurement of the medial longitudinal arch. *Arch Phys Med Rehabil.*, 1995, 76, 45–49.

[49] Saltzman, CL; Brandser, EA; Berbaum, KS; DeGnore, L; Holmes, JR; Katcherian, DA; Teasdall, RD; Alexander, IJ. Reliability of standard foot radiographic measurements. *Foot Ankle Int.*, 1994, 15, 661–665.

[50] Sawacha, Z. Validation of plantar pressure measurements for a novel in-shoe plantar sensory replacement unit. *J Diabetes Sci Technol.*, 2013, Sep 1, 7(5), 1176-8.

[51] Schmiegel, A; Rosenbaum, D; Schorat, A; Hilker, A; Gaubitz, M. Assesment of foot impairment in rheumatoid arthritis patients by dynamic pedobarography. *Gait Posture*, 27(1), 100-104, 2008.

[52] Schuh, R; Gruber, F; Wanivenhaus, A; Hartig, N; Windhager, R; Trnka, HJ. Flexor digitorumlongus transfer and medial displacement calcaneal osteotomy for the treatment of stage II posterior tibial tendon dysfunction: kinematic and functional results of fifty one feet. *IntOrthop.*, 37(9), 1815-1820, 2013.

[53] SooHoo, NF; Zingmond, DS; Ko, CY. Comparison of reoperation rates following ankle arthrodesis and total ankle arthroplasty. *J. Bone Joint Surg. Am.*, 2007, 89 (10), 2143–2149, (Oct).

[54] Stengel, D; Bauwens, K; Ekkernkamp, A; Cramer, J. Efficacy of total ankle replacement with meniscal-bearing devices: a systematic review and meta-analysis. *Arch. Orthop. Trauma Surg.*, 2005, 125 (2), 109–119, (Mar).

[55] Tenenbaum, S; Coleman, SC; Brodsky, JW. Improvement in gait following combined ankle and subtalar arthrodesis. *J. Bone Joint Surg. Am.*, 2014, 96 (22), 1863–1869 (Nov 19).

[56] Thomas, R; Daniels, TR; Parker, K. Gait analysis and functional outcomes following ankle arthrodesis for isolated ankle arthritis. *J. Bone Joint Surg. Am.*, (2006), 88 (3), 526–535, (Mar).

[57] Veronese, N; Stubbs, B; Volpato, S; et al. Association Between Gait Speed With Mortality, Cardiovascular Disease and Cancer: A Systematic Review and Meta-analysis of Prospective Cohort Studies. *JAMDA*, 19, (2018), 981-988.

[58] Vuillerme, N; Chenu, O; Demongeot, J; Payan, Y. Controlling posture using a plantar pressure-based, tongue-placed tactile biofeedback system. *Exp Brain Res.*, 2007 May, 179(3), 409-14.

[59] Weseley, MS; Koval, R; Kleiger, B. Roentgen measurement of ankle flexion-extension motion. *Clin Orthop Relat Res.*, 1969, 65, 167–174.

[60] Zanolli, DH; Nunley, JA; 2nd. Easley, ME. Subtalar fusion rate in patients with previous ipsilateral ankle arthrodesis. *Foot Ankle Int.*, 2015 Sep, 36(9), 1025-8.

[61] Ziegler, P; Friederichs, J; Hungerer, S. Fusion of the subtalar joint for post-traumatic arthrosis: a study of functional outcomes and non-unions. *Int Orthop.*, 2017 Jul, 41(7), 1387-1393.

[62] Wood, PL; Sutton, C; Mishra, V; Suneja, R. A randomized, controlled trial of two mobile-bearing total ankle replacements. *J. Bone Joint Surg. (Br.)*, (2009), 91 (1), 69–74 (Jan).

[63] Zong-Hao, MaC; Hong-Ping, Wan A; Wai-Chi Wong, D; et al. A vibrotactile and plantar force measurement-based biofeedback system: paving the way towards warable balance-improving devices. *Sensors*, 2015, 15, 31709-31722.

INDEX

A

acromion, x, 56, 62, 67, 105
acromion height asymmetry, 67
adaptation, ix, 20, 84
adjustment, 44, 85
age, 4, 7, 60, 61, 91, 129, 152
anatomy, 34, 80
anchorage, viii, 2, 4, 5, 7, 11, 12, 16, 17
ankle arthrodesis, xii, 134, 136, 139, 146, 149, 150, 151, 154, 155, 156, 158, 159
anthropometry discrepancy, 56, 66
arthritis, 147, 154, 158
arthrodesis, xii, 36, 52, 134, 136, 139, 146, 147, 149, 150, 151, 153, 154, 155, 156, 157, 158, 159
arthroplasty, 153, 155, 158
asymmetry, vii, x, 55, 59, 60, 61, 65, 66, 68, 74, 76, 77, 78, 79, 80, 82, 83, 86, 88, 89, 92
asymmetry on hip and lower back, 83
asymmetry on trochanteric height, 83
athletes, 100, 101, 106, 107, 108, 110, 111, 116, 123, 124, 127
augmentation, v, viii, 1, 2, 3, 4, 5, 6, 7, 8, 9, 10, 11, 12, 13, 15, 16, 17, 48

B

bilateral, x, 56, 59, 60, 61, 62, 73, 83, 84
biofeedback, 145, 151, 152, 154, 157, 158, 159
biomaterials, xi, 99, 101
biomechanics, vii, xi, 14, 17, 19, 20, 33, 34, 35, 39, 40, 50, 87, 92, 97, 99, 100, 106, 127, 128, 129, 131, 132, 156
bipedal, 57, 58, 59, 76, 83, 87, 88, 93
bipedalism, x, 55, 57, 75
biphasic, xii, 134, 136, 141, 143
blood, 80, 86, 110
blood circulation, 80, 86
Bluetooth, xiii, 134, 137, 138
body image, 104, 105, 120
body morphology, xi, 99, 100, 132
body weight, 77, 78, 140, 142, 152
bone, vii, viii, ix, 2, 3, 4, 5, 6, 10, 11, 12, 13, 14, 16, 19, 20, 21, 22, 23, 24, 25, 28, 31, 33, 34, 38, 40, 47, 54, 79, 80, 91, 106, 107, 152, 157
bone resorption, 31, 47
brain, 75, 76

Index

C

cables, xiii, 134, 136, 138
cartilage, 146, 152
CCF coefficient, xi, 56, 65, 66, 69, 70, 71, 72, 73, 74, 77, 79, 81, 82
central pattern generators (CPG), 84
cervical laminectomy, 36, 51
children, 60, 78, 91
chimpanzee, 83, 87
clinical application, vi, 12, 133, 134, 136, 145
clinical outcome, 146
commercial, xiii, 134, 136
competition, 100, 101, 107, 124
competitors, 100, 110, 129, 132
complications, 37, 41, 53, 156
compression, 4, 6, 30, 42, 46
computer, 21, 30, 33, 45, 63, 105, 152
configuration, ix, 17, 20, 25, 29, 31, 32, 40, 41, 42, 44, 45, 47, 111
construction, xi, 99, 101, 102
contact times, xii, 134, 136, 140, 141, 142
contralateral feet, xii, 134, 136, 144, 155
control group, xii, 134, 136
coordination, 56, 60, 81, 82, 83, 85, 88, 90
corpectomy, viii, ix, 19, 20, 21, 23, 25, 29, 31, 37, 38, 40, 41, 53, 54
correlation, x, 12, 55, 56, 65, 66, 74, 78, 79, 82, 86, 91, 139, 141, 144, 150
correlation analysis, 65, 79
correlation coefficient, 66, 144, 150
correlation function, x, 56, 65
cortical bone, 21, 24
cross-correlation function, 56, 65, 69, 72
cycling, 109, 152

D

data analysis, 8, 9, 86
deformation, xi, 99

depression, xii, 134, 136, 141, 143, 149
Diabetes, 154, 158
diabetic ulcerations, 136, 145
displacement, 33, 42, 43, 44, 45, 59, 158
distribution, 89, 101, 121, 132, 139, 145, 156, 157
dominance, 59, 60, 75, 83, 87, 88, 92
dynamic plate, ix, 20, 24, 25, 28, 30, 31, 32, 37, 38, 40, 41, 43, 44, 45, 46, 47, 49, 52, 53, 54

E

elastoplasty, 2, 17, 18
electromyogram/electromyography (EMG), 56, 59, 62, 63, 65
energy, 62, 63, 82, 83, 86, 116, 151
engineering, 100, 102, 109, 129
equilibrium, 85, 100
equipment, xi, xii, 100, 120, 133, 135, 137, 138, 151
erector spinae, x, 56, 62, 73, 74, 82, 83, 84, 87
evolution, x, 55, 75, 85, 87
exoskeleton, 107, 109
extrusion, ix, 20, 40

F

femur, 107, 123
fibula, ix, 20, 40
fixation, 16, 33, 34, 38, 40, 49, 53, 54, 157
flexor, 63, 80, 81, 91
foot pressure, xi, 56, 60, 63, 64, 65, 68, 69, 74, 77, 78, 79, 83, 85, 91, 92, 154
footedness, x, 55, 57, 58, 61, 74, 77, 87, 88, 89, 92, 93
footwear, 63, 155
force, xii, 44, 59, 60, 63, 81, 86, 89, 91, 103, 107, 111, 112, 120, 121, 134, 136, 139, 141, 142, 143, 144, 147, 149, 153, 159

force-time curve, xii, 134, 136, 141, 144
fossils, 75, 76
fractures, viii, 1, 3, 7, 10, 11, 13, 14, 17, 90, 142, 146
friction, 24, 44, 64
fusion, viii, ix, 19, 20, 21, 25, 29, 30, 31, 33, 34, 35, 36, 37, 40, 41, 46, 48, 50, 51, 52, 53, 147, 149, 153, 155, 159

G

gait, xi, xii, 56, 57, 58, 59, 60, 61, 63, 64, 67, 68, 71, 72, 77, 78, 81, 82, 83, 86, 88, 89, 90, 91, 92, 93, 97, 134, 135, 136, 138, 141, 142, 143, 146, 147, 150, 153, 155, 156, 157, 158
gait cycle, xi, 56, 57, 58, 63, 64, 67, 68, 71, 72, 78, 92, 97
gait cycle duration, 67
geometrical parameters, 35, 51
geometry, vi, xi, 24, 32, 33, 35, 45, 47, 51, 99, 100, 101, 102, 103, 105, 106, 109, 115, 127, 128, 129, 132
gravity, 63, 82, 112, 128
growth, 82, 152
gymnastics, 107, 108

H

handedness, x, 55, 57, 65, 70, 71, 73, 74, 75, 76, 77, 80, 82, 84, 86, 87, 88, 89, 90, 92, 93
healing, 31, 41, 47, 146
healthy feet, xii, 134, 136, 139, 142, 144
height, x, 4, 31, 46, 56, 61, 62, 67, 74, 83, 105, 106, 108, 116, 117, 125, 126
hindfoot alignment, xiii, 134, 136, 151, 155
Hindfoot Alignment Guide, 151
hindfoot position, 150, 151
hip joint, 107, 126

human, xi, 21, 29, 32, 33, 34, 35, 36, 45, 50, 51, 52, 57, 59, 60, 61, 62, 75, 76, 78, 83, 84, 85, 86, 88, 89, 90, 91, 93, 99, 100, 101, 102, 106, 109, 114, 115, 116, 128, 129, 130, 131, 132
human body, xi, 57, 60, 78, 83, 85, 99, 100, 101, 102, 106, 109, 114, 115, 128, 129, 130, 132
human body symmetry, 60

I

iliac crest, ix, 20, 40
implants, 17, 23, 29, 45, 46, 152
in vitro, viii, ix, x, 2, 3, 4, 11, 15, 20, 40, 41, 49
in vivo, ix, x, 12, 20, 40
inertia, vi, vii, xi, 99, 100, 101, 111, 112, 113, 114, 120, 121, 122, 123, 126, 127, 128, 129, 132
injury, 35, 50, 145, 146, 152
integrated EMG (iEMG), 65
integration, ix, 20, 21, 26, 40, 41
interface, viii, xiii, 2, 3, 14, 17, 134, 136, 137, 138, 151
inversion, 63, 150
ipsilateral, 63, 82, 83, 159

J

joints, 24, 34, 50, 101, 102, 120, 150, 152
jumping, 124, 125

L

laterality, vi, vii, x, 55, 57, 59, 60, 61, 65, 72, 75, 76, 78, 82, 84, 85, 88, 89, 90, 92
laterality questionnaires, 61
left-handedness, 76, 80, 82, 85, 87

left-handers, x, 55, 58, 61, 62, 65, 66, 67, 68, 69, 70, 71, 72, 73, 76, 77, 79, 80, 81, 82, 83, 84, 85, 86
legs, 57, 80, 83, 106, 126
life expectancy, xi, 56, 86
localization, vii, xi, 100, 117, 120
locomotor, 87, 88
lordosis, 31, 46, 102
lower thigh asymmetry, 74
lower thigh circumference asymmetry, 74
lumbar spine, 33, 34

M

mass, vii, xi, 59, 78, 82, 100, 101, 103, 105, 112, 113, 114, 115, 116, 117, 118, 119, 120, 121, 122, 123, 124, 125, 126, 128, 129, 130, 132
materials, viii, 2, 4, 5, 6, 9, 11, 12, 14, 23, 41
maximal force, xii, 134, 136, 141, 142, 147
mechanical properties, viii, 2, 3, 11, 21, 29, 33, 45
meta-analysis, 156, 158
metatarsal, 63, 69, 70, 71, 72, 74, 77, 78, 79, 83, 139
midfoot depression, xii, 134, 136, 141, 143, 149
morphology, xi, 6, 21, 99, 100, 101, 127, 132
muscle strength, 83, 111
muscles, 60, 62, 64, 80, 82, 84, 87, 91, 107, 126
musculoskeletal, xi, 17, 56, 82, 85, 86

N

necrosis, 3, 14, 80
nodes, 21, 23, 24
nonunions, 152
nuclear magnetic resonance, xi, 99

O

organs, 100, 105, 110
osteoarthritis, xiii, 134, 136, 139, 142, 143, 146, 147, 150
osteoporosis, 23, 152
osteotomy, 151, 158

P

pain, 144, 146, 153
palpation, 62, 91
parameters, viii, xii, 4, 13, 26, 35, 51, 59, 93, 133, 135, 136, 140, 141, 142, 144, 147, 151, 156
pedicle screw anchorage, viii, 2, 4, 5, 11, 12, 16
pedobarography, vi, xii, 133, 134, 135, 136, 137, 138, 139, 140, 141, 145, 146, 150, 155, 158
pelvis, 63, 80, 105
peripheral neuropathy, 154, 157
physiology, 85, 86, 127
polymethylmethacrylate (PMMA), viii, 1, 2, 3, 4, 5, 7, 8, 9, 10, 11, 12, 13, 14, 15, 16, 17
polymer, vii, 2, 3, 11, 16, 17
polymerization, 13, 15
population, x, 55, 58, 61, 75, 76, 80, 85, 86, 102
positive correlation, 65, 73, 74, 78, 79, 144
preschool children, 78, 91
pressure sensor, x, 56, 58, 63, 64
primary school, 60, 91
prostheses, 35, 51, 101, 107, 108, 109

R

radius, 101, 112, 113, 120, 121, 122
radius of gyration, 121, 122

reconstruction, viii, ix, 19, 33, 40
relative midfoot index, viii, xii, 134, 136, 141, 142, 155
researchers, 59, 85, 95, 115, 145
resistance, 101, 102, 103, 111, 120
right-handedness, 75
right-handers, x, 56, 58, 61, 62, 65, 66, 67, 68, 69, 70, 71, 72, 73, 77, 79, 80, 81, 82, 83, 84, 85

S

safety, xi, 12, 56, 86
scholarship, 86, 95
sensitivity, 32, 51, 142
sensors, x, xiii, 56, 58, 63, 64, 68, 77, 134, 136, 138, 152
shape, 83, 101, 103, 104, 113
shoes, xii, 134, 136, 137, 147
silicone based elastomer, v, viii, 1, 2, 3, 10, 12
skin, 80, 102, 139
soleus, x, 56, 62, 72, 73, 74, 75, 81, 82, 83, 84
spatial symmetry, 57
spatio-temporal symmetry, 57
speed, x, 6, 42, 55, 60, 64, 67, 68, 69, 70, 72, 73, 74, 79, 81, 84, 85, 90, 91, 110, 111, 139, 140, 142, 146, 156, 158
spine, 3, 15, 16, 21, 22, 23, 24, 26, 29, 30, 32, 33, 34, 35, 36, 37, 38, 46, 49, 50, 51, 52, 53, 54
sport, vi, vii, xii, 76, 99, 100, 102, 106, 107, 110, 123, 129, 130, 131, 132, 135
sprains, 145, 146
stability, 31, 34, 47, 49, 54, 59, 60, 81, 82, 83, 90
stabilization, 2, 20, 34, 36, 37, 38, 52, 53, 54
stance phase, 57, 68, 77, 79, 80, 90, 138
standardization, xii, 134, 135, 136, 140

standardized reporting, 134, 140
stress, viii, 2, 30, 33, 41, 46, 83
stroke, 59, 89
structure, viii, xi, 2, 3, 11, 23, 26, 83, 99, 101
swing phase, 82, 138
symmetry, vii, x, 55, 56, 57, 59, 60, 65, 69, 70, 72, 81, 82, 83, 84, 85, 88, 91, 92, 95, 97
symmetry breaking, 56, 60
symmetry index, 59

T

techniques, viii, 2, 5, 6, 11, 12, 15, 36, 37, 49, 53, 85, 151, 152
tendon, 144, 152, 158
tension, 6, 28, 29, 44, 64
thorax, 104, 105
tibialis anterior, x, 56, 62, 80, 81
tibio-talo-calcaneal arthrodesis, 139
time lag, xi, 56, 65, 66, 69, 70, 71, 72, 73, 74, 77, 79, 80, 81, 82
tissue, 3, 14, 23, 33, 105
torsion, 26, 27
total ankle replacement, xii, 134, 136, 139, 146, 149, 151, 154, 155, 156, 158, 159
training, 13, 100, 127
translation, xiii, 30, 31, 46, 47, 134, 136
transmission, vii, ix, 20, 25, 33, 40, 45, 47
treatment, viii, 1, 10, 13, 37, 53, 83, 135, 146, 147, 152, 157, 158
trial, 37, 52, 63, 64, 65, 81, 159
triphasic, xii, 134, 136, 141, 143

U

uniform, 30, 46
unloading, 136, 145, 146, 152, 153, 154

V

valgus, xiii, 134, 136, 138, 150
variations, 35, 44, 50, 108
varus, xiii, 134, 135, 136, 138, 150
velocity, 103, 110, 126, 127, 141
vertebrae, viii, 2, 4, 7, 8, 9, 11, 14, 20, 34, 41
vertebroplasty, viii, 2, 3, 4, 7, 11, 13, 14, 15, 16, 17, 18
vision, 64, 91
visual judgment of hindfoot alignment, 150
VK100, 2, 4, 5, 7, 8, 9, 10, 11, 12, 13, 17
volleyball, 107, 108

W

walking, vii, x, 55, 57, 59, 60, 61, 62, 64, 65, 66, 67, 68, 69, 70, 71, 72, 73, 74, 77, 79, 81, 82, 83, 84, 85, 87, 88, 89, 90, 91, 92, 93, 97, 137, 139, 140, 142, 146, 153, 156, 157
walking asymmetry, vi, vii, x, 55, 60, 74
water, 102, 103, 107, 111
Waterloo Footedness Questionnaire (WFQ), 61, 66, 77, 78
Waterloo Handedness Questionnaire (WHQ), 61, 66, 78
wear, xii, 134, 135, 136, 137

X

x-rays, 105, 116

Related Nova Publications

Old Problems and New Horizons in World Physics

Editors: Volodymyr Krasnoholovets, PhD, Victor Christianto, and Florentin Smarandache, PhD

Series: Physics Research and Technology

Book Description: Written by 13 contributors from different regions of the World, this book is a collection of papers written by researchers who have been working toward defining new concepts in the sciences for years. Among the new approaches, new views have been developed based on the emerging mathematical principles, the observation of possible relationships between physical processes, and ideas inspired by firsthand experience penetrating elusive realms.

Hardcover ISBN: 978-1-53615-430-6
Retail Price: $230

Magnifying Spacetime: How Physics Changes with Scale

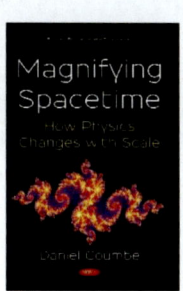

Author: Daniel Coumbe

Series: Physics Research and Technology

Book Description: *Magnifying Spacetime* delivers new insights into the role of scale in quantum gravity from the cutting-edge of modern research using an accessible and pedagogical style. The ideal complementary text for undergraduate and graduate students, this book also serves as an essential resource for professional physicists working on related topics.

Softcover ISBN: 978-1-53615-319-4
Retail Price: $82

To see a complete list of Nova publications, please visit our website at www.novapublishers.com

Related Nova Publications

AN ESSENTIAL GUIDE TO ELECTRICAL CONDUCTIVITY AND RESISTIVITY

EDITOR: Luke Lewin

SERIES: Physics Research and Technology

BOOK DESCRIPTION: An Essential Guide to Electrical Conductivity and Resistivity opens with experimental and theoretical data on the important structurally sensitive property of the molten oxide-chloride systems KCl (50 mol. %)–PbCl2 (50 mol. %), CsCl (18.3 mol. %)–PbCl2 (81.7 mol. %) and CsCl (71.3 mol. %)–PbCl2 (28.7 mol. %) with PbO concentration reaching 20 mol.% in the temperature range of 764 – 917 K.

SOFTCOVER ISBN: 978-1-53615-047-6
RETAIL PRICE: $95

AN ESSENTIAL GUIDE TO ELECTRODYNAMICS

EDITOR: Norma Brewer

SERIES: Physics Research and Technology

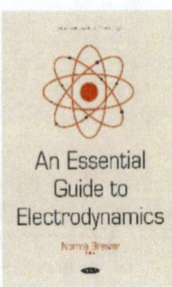

BOOK DESCRIPTION: The opening chapter of *An Essential Guide to Electrodynamics* describes a new theory of the electron, from which derives a fully deductive explanation of the chemical inertness of the group 18 elements of the periodic system.

HARDCOVER ISBN: 978-1-53615-705-5
RETAIL PRICE: $230

To see a complete list of Nova publications, please visit our website at www.novapublishers.com